METHODS IN MEDICAL INFORMATICS

Fundamentals of Healthcare Programming
in Perl, Python, and Ruby

CHAPMAN & HALL/CRC
Mathematical and Computational Biology Series

Aims and scope:

This series aims to capture new developments and summarize what is known over the entire spectrum of mathematical and computational biology and medicine. It seeks to encourage the integration of mathematical, statistical, and computational methods into biology by publishing a broad range of textbooks, reference works, and handbooks. The titles included in the series are meant to appeal to students, researchers, and professionals in the mathematical, statistical and computational sciences, fundamental biology and bioengineering, as well as interdisciplinary researchers involved in the field. The inclusion of concrete examples and applications, and programming techniques and examples, is highly encouraged.

Series Editors

N. F. Britton
Department of Mathematical Sciences
University of Bath

Xihong Lin
Department of Biostatistics
Harvard University

Hershel M. Safer

Maria Victoria Schneider
European Bioinformatics Institute

Mona Singh
Department of Computer Science
Princeton University

Anna Tramontano
Department of Biochemical Sciences
University of Rome La Sapienza

Proposals for the series should be submitted to one of the series editors above or directly to:
CRC Press, Taylor & Francis Group
4th, Floor, Albert House
1-4 Singer Street
London EC2A 4BQ
UK

Chapman & Hall/CRC Mathematical and Computational Biology Series

METHODS IN MEDICAL INFORMATICS

Fundamentals of Healthcare Programming in Perl, Python, and Ruby

Jules J. Berman

CRC Press
Taylor & Francis Group
Boca Raton London New York

CRC Press is an imprint of the
Taylor & Francis Group, an **informa** business

A CHAPMAN & HALL BOOK

CRC Press
Taylor & Francis Group
6000 Broken Sound Parkway NW, Suite 300
Boca Raton, FL 33487-2742

First issued in paperback 2018

© 2011 by Taylor and Francis Group, LLC
CRC Press is an imprint of Taylor & Francis Group, an Informa business

No claim to original U.S. Government works

ISBN-13: 978-1-4398-4182-2 (hbk)
ISBN-13: 978-1-138-37441-6 (pbk)

Library of Congress Cataloging-in-Publication Data

Berman, Jules J.
    Methods in medical informatics : fundamentals of healthcare programming in Perl, Python, and Ruby / Jules J. Berman.
        p. ; cm. -- (Chapman & Hall/CRC mathematical and computational biology series ; 39)
    Includes bibliographical references and index.
    ISBN 978-1-4398-4182-2 (alk. paper)
    1. Medical informatics--Methodology. 2. Medicine--Data processing. I. Title. II. Series: Chapman and Hall/CRC mathematical & computational biology series ; 39.
    [DNLM: 1. Medical Informatics--methods. 2. Programming Languages. 3. Computing Methodologies. W 26.5 B516m 2011]

R858.B4719 2011
610.285--dc22

2010011244

Visit the Taylor & Francis Web site at
http://www.taylorandfrancis.com

and the CRC Press Web site at
http://www.crcpress.com

For Irene

Contents

PART III PRIMARY TASKS OF MEDICAL INFORMATICS

Preface

There are many talented and energetic healthcare workers who have basic programming skills, but who have not had an opportunity to use their skills to help their patients or advance medical science. Too often, healthcare workers are led to believe that medical informatics is a complex and specialized field that can only be mastered by teams of professional programmers. This is just not the case. A few dozen simple algorithms account for the bulk of activities in the field of medical informatics. Moreover, in the past decade, gigabytes of medical data, comprising many millions of deidentified clinical records, have been released into the public domain, and are freely accessible via the Internet. With the arrival of open source high-level programming languages, the barriers to entry into the field of medical informatics have collapsed.

Innovative medical data analysis cannot be driven by commercial software applications. There are limits to what anyone can accomplish with spreadsheets, statistical packages, search engines, and other off-the-shelf computational products. There will come a point, in the careers of all healthcare professionals, when they need to perform their own programming to answer a very specific question, or to discover a new hypothesis from a trove of data resources. This book provides step-by-step instructions for applying basic informatics algorithms to medical data sets. It is written for students and professionals in the healthcare field who have some working knowledge of Perl, Python, or Ruby. Most of our future data analysis efforts will build on the computational approaches and programming routines developed in this book.

Perl, Python, and Ruby are free, readily available, open source programming languages that can be used on any operating system including Windows, Linux, and Mac. Most people who work in the biomedical sciences and develop their own programming solutions, perform at least some of their programming with one of these three languages. These languages are popular, in part because they are easy to learn.

Without becoming a full-time programmer, you can write powerful programs, in just a few minutes and a few lines of code, with any of these languages.

We will use a minimal selection of commands to write short scripts that can be learned quickly by biomedical students and professionals. This book demonstrates that, with a few programming methods, biomedical professionals can master any kind of data collection.

Though there are numerous books that introduce programming techniques to biomedical professionals (including several that I have written) no other book has these important features:

1. All of the data, nomenclatures, programming scripts, and programming languages used in this book are free and publicly available. Most of the data comes from U.S. government sources, providing gigabytes of high quality, curated biomedical data to a global community of scientists, healthcare experts, clinicians, nurses, and students. Every student should become familiar with these data sources, and understand their medical value. This book provides instructions for downloading all of the data sources discussed in the book.

2. Data come in many different forms. We describe the structure of every data source used. In the case of image formats, we provide instructions for converting between the different file types.

3. Most medical informatics books are written for one specific language, or are written as "concept books" that describe algorithms without actually providing programming instruction. We provide equivalent scripts in Perl, Python, and Ruby, so that anyone with some programming skill will benefit. Each trio of scripts is preceded by a step-by-step explanation of the algorithm, in plain English. You may wish to confine your attention to scripts written in your preferred language. Over the years, you may find it valuable to reread this book, paying attention to the languages you ignored on the first pass.

4. It is nearly impossible to begin a new data analysis project without first observing some case examples. With step-by-step instructions, you will learn the basic informatics methods for retrieving, organizing, merging, and analyzing the following data sources.

Here are the public resources used in this book:

Data Sets and Services

SEER—The National Cancer Institute's Surveillance Epidemiology and End Results project, containing deidentified records for nearly 4 million cancer cases.

PubMed—The National Library of Medicine's Web-based bibliographic retrieval service. The title, author(s), journal publication information, and, in most cases, article summaries, are provided for over 19 million medical citations.

CDC mortality data sets—The Centers for Disease Control and Prevention's collection of mortality records containing computer-parsable data on virtually every death occurring in the U.S.

U.S. Census—Every 10 years, the U.S. Bureau of Census counts the number of people living in the U.S., and collects basic demographic information in the process. Much of the information collected by the census is freely available to the public.

OMIM®—The Online Mendelian Inheritance in Man® is a large data set containing detailed information on over 20,000 inherited conditions of humans, made publicly available by the National Library of Medicine's National Center for Biotechnology Information.

Nomenclatures and Ontologies

MeSH—Medical Subject Headings, a comprehensive, hierarchical listing of medical topics, developed by the National Library of Medicine.

ICD and ICD-O—The World Health Organization's disease nomenclatures, the International Classification of Diseases and the International Classification of Diseases in Oncology.

Taxonomy—A computer-parsable classification of organisms, used by biotechnology centers.

Developmental Lineage Classification and Taxonomy of Neoplasms—The largest nomenclature of tumors in existence, with synonymous terms grouped under concepts and organized as a hierarchical biological classification.

Internet Protocols, Markup Languages, and Interfaces

HTML—HyperText Markup Language, the markup language used in Web pages.

HTTP—Hypertext Transfer Protocol, the Internet protocol supporting the Internet's World Wide Web.

XML—eXtensible Markup Language, a syntax for describing the data and including both data and data descriptors in a format that can be read by humans and computers.

RDF—Resource Description Framework, a method of organizing information in statements that bind data, and descriptors for the data, to an identified object. RDF is expressed in the XML markup language.

CGI—Common Gateway Interface, an Internet protocol, used by Perl, Python, Ruby, and other languages, that receives input values submitted through Web pages.

The included scripts will call upon a few programming skills, in either Perl, Python, or Ruby. You should know the basic syntax of a language, the minimum structural requirements for a script, how command lines are written, how iterating loops are structured, how files are opened, read, and written, how values can be assigned to and retrieved from data structures, how simple regular expressions are interpreted, and how scripts are launched. The scripts are written in a style that sacrifices elegance for readability. If your knowledge of Perl, Python, or Ruby is shaky, there are numerous beginner-level books, and many Web-based tutorials for each of these languages.

The book is divided into four parts: Part I—Fundamental Algorithms and Methods of Medical Informatics; Part II—Medical Data Resources; Part III—Primary Tasks of Medical Informatics; and Part IV—Medical Discovery.

Part I—Fundamental Algorithms and Methods of Medical Informatics (Chapters 1 to 4) provides simple methods for viewing text and image files, and for parsing through large data sets line by line, retrieving, counting, and indexing selected items. The primary purpose of these chapters is to introduce the basic computational subroutines that are used in more complex scripts later in the book. The secondary purpose of these chapters is to demonstrate that Perl, Python, and Ruby are quite similar to one another, and provide equivalent functionality.

Part II—Medical Data Resources (Chapters 5 to 13) demonstrates uses of some freely available biomedical data sets. These data sets have cost hundreds of millions of dollars to assemble, yet many healthcare workers are unaware of their enormous clinical value. In these chapters, you will learn the intended uses of data sets, how the data sets are organized, and how you can select, retrieve, and analyze information from the files.

Part III—Primary Tasks of Medical Informatics (Chapters 14 to 18) covers some of the computational methods of biomedical informatics, including autocoding, data scrubbing, and data deidentification.

A good question is hard to find. Part IV—Medical Discovery (Chapters 19 through 27) provides examples of the kinds of questions that biomedical scientists can ask and answer with public data and open source programming languages. In these chapters, we combine methods developed in the earlier chapters, using freely available data sources to answer specific questions or to develop new medical hypotheses. Many of the informatics projects that you will use in your biomedical career can be completed with the basic methods and implementations described in these chapters.

This book is intended to be used as a textbook in medical informatics courses. Because the methods in the book are generalized, the book will also serve as a convenient reference source of script snippets that can be freely used by students and professionals. The scripts are written in a syntax appropriate for the most current popular version of Perl, Python, or Ruby, and based on the availability of about a dozen large, public data sets, each with a consistent data structure. Over time, programming languages change; the availability, Internet location, and organization of the large public data sets may also change. Readers should be warned that, as time goes by, the scripts will need to be modified. Because the scripts are very short, future script changes should be minor, and easy to implement.

I maintain a Web site with updated resources for all of my books (including this one) at the following address: http://www.julesberman.info/.

Nota Bene

Throughout the book are short scripts. Most of the scripts are under a dozen lines of code, and every script is preceded by a step-by-step explanation of the code's basic algorithm. To keep the scripts short, easy to understand, and generalizable, I omitted many of the tricks and language-specific conventions that programmers love to flaunt: subroutines, pragmas, exception handling, references, nested data structures, command-line parameters, and iterator functions (to name a few). Every script was tested and successfully executed in the Windows® operating system, using Perl version 5.8, Python version 2.5, and Ruby version 1.8. Because the scripts are all short and simple, using a minimum of external modules, it is likely that many of the scripts will execute without modification, on any computer. Some scripts will require publicly available data files that you must download to your own computer. You will need to modify these scripts to include the correct directory locations for your own file system. An archive of small text and image files, used throughout the book, along with all of the book scripts, are available from the publisher's Web site. Please note that a return arrow, shown at right, indicates a line continuation and is not script code.

The following disclaimer applies to all parts of this book, including text, scripts, and images. This material is provided by its creator, Jules J. Berman, "as is," without warranty of any kind, expressed or implied, including but not limited to the warranties of merchantability, fitness for a particular purpose, and noninfringement. In no event shall the author or copyright holder be liable for any claim, damages, or other liability, whether in an action of contract, tort or otherwise, arising from, out of, or in connection with the material or the use or other dealings.

All of the scripts included in the book are distributed under the GNU General Public License, a copy of which is available at

http://www.gnu.org/copyleft/gpl.html

If you encounter problems with the scripts, the author will try to find the problem and make corrections. The author cannot guarantee that a correction or modification will satisfy the needs or the desires of every reader. Readers should understand that this book is a work of literature, and not a collection of software applications.

About the Author

Jules Berman, Ph.D., M.D., received two bachelor of science degrees (mathematics and earth sciences) from MIT, a Ph.D. in pathology from Temple University, and an M.D. from the University of Miami School of Medicine. His postdoctoral research was conducted at the National Cancer Institute. His medical residence in pathology was completed at the George Washington University School of Medicine. He became board certified in anatomic pathology and in cytopathology, and served as the chief of Anatomic Pathology, Surgical Pathology and Cytopathology at the Veterans Administration (VA) Medical Center in Baltimore, Maryland. While at the Baltimore VA, he held appointments at the University of Maryland Medical Center and at the Johns Hopkins Medical Institutions. In 1998, he became the program director for pathology informatics in the Cancer Diagnosis Program at the U.S. National Cancer Institute. In 2006, he became president of the Association for Pathology Informatics. Over the course of his career, he has written, as first author, more than 100 publications, including five books in the field of medical informatics. Today, Dr. Berman is a full-time freelance writer.

PART I

FUNDAMENTAL ALGORITHMS AND METHODS OF MEDICAL INFORMATICS

1

PARSING AND TRANSFORMING TEXT FILES

File parsing is the computational method of reading a file, piece by piece. The size of the pieces can be miniscule (individual bits) or small (individual characters), or somewhat bigger (lines of text, paragraphs, demarcated reports, etc.). In most parsing routines, the pieces are matched against a pattern, then extracted or modified. Modifications can include translations, substitutions, deletions, additions, or formatting changes. The bulk of biomedical informatics routines involve some element of file parsing.

1.1 Peeking into Large Files

Some of the files we will be using exceed a gigabyte in length. Most word processors simply cannot open a file of this size. You will need a simple utility that can open a large file, extract a sample of the file, and display it on your monitor. In a few lines of code, we can write a script that will extract and display the first 40 lines from a large text file, and will store the first 3,000 lines in a separate file that you can open with your word processor.

1.1.1 Script Algorithm

1. Send a prompt to the monitor asking for the name of a file to be searched.
2. Receive the line of text entered by the user, clipping off the carriage return (also known as the *newline character*) that is always added when the user pushes the Enter key.
3. Put the received keyboarded response into a variable that contains the name of the file to be searched.
4. Open the file for reading.
5. Open another file for writing. This file will receive the output of the script.
6. Create a "for" loop that will iterate 40 times.
7. In each loop, read a line of text. Print the line of text to the monitor, and print the same line of text to the "write" file you opened in step 5.
8. Create a "for" loop that will iterate 2,960 times. This loop will continue reading the large file at the location where the prior (40 iteration loop) stops.
9. In each loop, read a line of text and send it to the "write" file opened in step 5.

10. When the loop is finished, print the name of the "write" file to the monitor (so that the user will know where to find the output text).
11. Exit. The opened files will automatically close when the script exits.

Perl Script

```perl
#!/usr/local/bin/perl
print "What file would you like to sample?\n";
$lookup = <STDIN>;
chomp();
open (TEXT, $lookup)||die"Can't open file";
open (OUT, ">sample.txt")||die"Can't open file";
$line = " ";
$count = 0;
print "\n";
for (0..39)
    {
     $line = <TEXT>;
     print OUT $line;
     print $line;
    }
for (0..2959)
    {
    $line = <TEXT>;
    print OUT $line;
    }
print "\nYour sampled text is in file \"sample.txt\"\n";
exit;
```

Python Script

```python
#!/usr/local/bin/python
import sys
import string
print "What file would you like to sample?\n"
line = sys.stdin.readline()
line = line.rstrip()
infile = open (line, "r")
outfile = open ("sample.txt", "w")
print "\n"
for iterations in range(40):
   getline = infile.readline()
   print getline.rstrip()
   outfile.write(getline)
for iterations in range(2960):
   getline = infile.readline()
   outfile.write(getline)
infile.close()
outfile.close()
```

```
print "\nYour sampled text is in file \"sample.txt\"\n"
exit
```

Ruby Script

```
#!/usr/local/bin/ruby
puts "What file would you like to read?";
filename = gets.chomp
file_in = File.open(filename,"r")
file_out = File.open("sample.txt","w")
(1..40).each {|n| puts file_in.readline}
(41..3000).each {|n| file_out.puts file_in.readline}
puts "Your sampled text is in file sample.txt"
exit
```

1.1.2 Analysis

Even simple scripts occasionally require the user to enter information via the keyboard. In this script, one command line is all that is needed to initiate a conversation between script and user. A line of text is sent to the monitor, and the script waits until the user enters a reply and presses the Enter key. The reply is captured by the script, and assigned to a script variable. Scripting languages provide a simple but effective user interface.

1.2 Paging through Large Text Files

Rather than snatching a portion of a large file (as in the prior example), you may prefer to read the file line by line until you tire of the process. Here is a script that displays the first 40 lines from any file, provides an opportunity to quit; if declined, displays the next 40 lines, and repeats indefinitely. By simply keeping your finger on the Enter key (thus bypassing the exit prompt), you can quickly scroll through the file. This script is particularly useful for large text files (exceeding 10 megabytes [MB]) that word processors cannot quickly load.

1.2.1 Script Algorithm

1. Send a prompt to the monitor asking for the name of a file that you want to read.
2. Receive the line of text entered by the user, clipping off the carriage return (newline character) that is always added when the user pushes the Enter key.
3. Put the received keyboarded response into a variable that now contains the name of the file to be searched.
4. Open the file.
5. Print the first 40 lines of the file.
6. Prompt the user, asking if he or she would like to quit the program.

7. If the user enters "QUIT" after the prompt, exit the program.

8. Otherwise, repeat steps 4, 5, and 6.

Perl Script

```
#!/usr/local/bin/perl
print "What file do you want to read?";
$filename = <STDIN>;
chomp($filename);
open (TEXT, $filename)||die"Can't open file";
$line = " ";
while ($line ne "")
    {
    for ($count = 1; $count <= 40; $count++)
     {
     $line = <TEXT>;
     print $line;
     }
    print "Type QUIT if you want to quit. Otherwise press any key\n";
    $response = <STDIN>;
    if ($response =~ /QUIT/i)
     {
     last;
     }
    }
exit;
```

Python Script

```
#!/usr/local/bin/python
import sys
import string
print "What file would you like to read?\n"
line = sys.stdin.readline()
line = line.rstrip()
infile = open (line, "r")
print "\n"
while(1):
    for iterations in range(40):
      print infile.readline().rstrip()
    print "Type QUIT if you want to quit. Otherwise press any key\n";
    response = sys.stdin.readline().rstrip()
    if (response == "QUIT"):
      break
infile.close()
exit
```

Ruby Script

```
#!/usr/local/bin/ruby
puts "What file would you like to read?";
filename = gets.chomp
```

```
file_in = File.open(filename,"r")
response = ""
count = 0
while 1
  file_in.each_line do
      |line|
      count = count + 1
      print line
      if count == 40
       count = 0
       puts "Type QUIT if you want to quit. Otherwise press any key.";
        response = gets.chomp
        exit if response == "QUIT"
        exit if response == "quit"
      end
  end
end
exit
```

1.2.2 Analysis

If you want to try this script, be sure to provide the name of a text file (a file consisting of standard ASCII characters) at the prompt, and give the full path to the file if it does not reside in the same subdirectory as your script.

Programming languages can open a file for reading, without loading the entire file into memory. When a file is opened for reading, file information can be accessed by sequential line readings, or by direct access to any selected byte location in the file. These operations are done very quickly. Regardless of the size of the file you want to access, each line will appear on your monitor before you can lift your finger from the Enter key. The rate limiting factor is the speed with which your monitor can display text.

1.3 Extracting Lines that Match a Regular Expression

Perl, Python, and Ruby support regular expression (regex) operations. Regex is a conventional way of describing string patterns.

An example of a regular expression is:

```
^[A-Z][a-z]+\s[0-9]*
```

This regex specifies the following pattern, "The string begins with an uppercase letter, followed by one or more lowercase letters followed by a space, followed by a succession of zero or more numerical digits."

With regex, you can search for classes of data. For example, an uppercase C followed by 7 numeric characters may specify a nomenclature code. A series of A,C,G,T characters may represent a gene sequence. A specific word or phrase, followed by as many as 50 characters of any value, followed by another specific word or phrase, may constitute a so-called proximity match (i.e., the relative co-location of two phrases). A set of alphabetic characters, forming a word, and beginning with a particular sequence of letters, may be the pattern that can pull every word with a common root.

Beyond pattern matching, regex is used for pattern substitution. Scripts can locate all the matches to a pattern and substitute another sequence of characters. The substitution sequence can be a specific word or character string, or it may be the product of an operation on the matched string (e.g., return the matched string as an all-uppercase string).

Regex is an extremely powerful tool for anyone working in the information field. Here is a basic regex script that parses through a file, extracting lines that contain a sequence that matches the user-provided regex pattern.

1.3.1 Script Algorithm

1. Send a prompt to the monitor asking for the name of a file to be searched.
2. Receive the line of text entered by the user and clip off the newline character.
3. Put the received keyboarded response into a variable that contains the name of the file to be searched.
4. Send a prompt to the monitor asking for a word, phrase, or regular expression to be searched.
5. Receive the line of text entered by the user, clipping off the newline character.
6. Put the received keyboarded response into a variable that contains the name of the regular expression that will be matched against every line in the file.
7. Open the file to be searched for reading.
8. Open a file, named "result.out", for writing. This file will hold your search results.
9. Parse through every line of the search file.
10. Whenever a line is encountered that matches your search expression, print it to the screen, and print it to the "result.out" file.
11. Exit. The opened files will automatically close when the script exits.

Perl Script

```
#!/usr/local/bin/perl
print "What file would you like to search?\n";
$filename = <STDIN>;
chomp($filename);
print "Enter a word, phrase or regular expression to search.\n";
$regex = <STDIN>;
chomp($regex);
open (TEXT, "$filename");
open (OUT, ">result.out");
while (<TEXT>)
```

```
      {
      $line = <TEXT>;
      if ($line =~ /$regex/i)
          {
          print OUT $line;
          print $line;
          }
      }
exit;
```

Python Script

```python
#!/usr/local/bin/python
import sys
import string
import re
print "What file would you like to search?";
filename = sys.stdin.readline()
filename = filename.rstrip()
print "Enter a word, phrase or regular expression to search."
regex = sys.stdin.readline()
regex = regex.rstrip()
infile = open (filename, "r")
outfile = open ("result.out", "w")
regex_object = re.compile(regex, re.I)
for line in infile:
    m= regex_objcct.scarch(line)
    if m:
        print line,
        outfile.write(line)
exit
```

Ruby Script

```ruby
#!/usr/local/bin/ruby
puts "What file would you like to search?";
filename = gets.chomp
puts "Enter a word, phrase or regular expression to search.";
regex = gets.chomp
f_in = File.open(filename,"r")
f_out = File.open("result.out", "w")
f_in.each do
  |line|
  if (line =~ /#{regex}/)
    print line
    f_out.print line
  end
end
exit
```

1.3.2 Analysis

When you try this script, be sure to provide the name of a text file (a file consisting of standard ASCII characters) at the prompt, and provide the full path to the file if it does not reside in the same subdirectory as your script. If you do not know how to compose a regular expression, just enter a search word or phrase at the prompt. The script will display every line from the provided file that contains a string that matches your search word or phrase, and will send a copy of the results to an external file.

1.4 Changing Every File in a Subdirectory

String substitution is a common computational task. Maybe you will want to switch every occurrence of the word "tumor" with "tumour" when submitting a manuscript to a British journal. Maybe a calculation, repeated throughout your quality assurance report, was incorrect; you want to substitute the correct number wherever the incorrect number appears. Maybe you have been spelling "Massachusetts" in a consistent, but incorrect manner.

The following script will parse through every file in a subdirectory, making a specific substitution at every matching sequence within every file.

1.4.1 Script Algorithm

1. Open a directory for reading. Do not run your script from the same directory as you are opening, because we will be modifying the files in the directory, and we do not want to modify our script while it is executing.
2. Put the list of files in the directory into an array.
3. Close the directory for reading.
4. Change the current directory to the directory that you previously opened.
5. For each file in your file list array, do the following: open the file, read through every line in the file, make the desired substitution for each matching sequence in each line, and close the file when you're finished.
6. After all of the files in the list have been parsed, exit.

Perl Script

```perl
#!/usr/local/bin/perl
opendir(FTPDIR, "c\:\\ftp\\some\\") || die ("Unable to open
directory");
@files = readdir(FTPDIR);
closedir(FTPDIR);
chdir("c\:\\ftp\\some\\");
undef($/);
foreach $file (@files)
  {
```

```
open (TEXT, $file);
$line = <TEXT>;
$line =~ s/\nexit\;/\nso long\;/g;
close TEXT;
unlink $file;
open (TEXT, ">$file");
print TEXT $line;
close TEXT;
}
exit;
```

Python Script

```
#!/usr/local/bin/python
import sys, os, re
filelist = os.listdir("c:\\ftp\\some\\")
os.chdir("c:\\ftp\\some\\")
for file in filelist:
  infile = open(file,'r')
  filestring = infile.read()
  infile.close()
  pattern = re.compile("exit")
  filestring = pattern.sub('so long', filestring)
  outfile = open(file,'w')
  outfile.write(filestring)
  outfile.close
exit
```

Ruby Script

```
#!/usr/local/bin/ruby
filelist = Dir.glob("c:/ftp/some/*.*")
filelist.each do
 |filepathname|
 contents = IO.read(filepathname)
 contents.gsub!("exit","so long") if contents.include? "exit"
 outfile = File.open(filepathname, "w")
 outfile.write(contents)
 outfile.close
end
exit
```

1.4.2 Analysis

Programming languages provide a simple way to determine the names of the files in a subdirectory. Once the names of the files are determined, it becomes straightforward to open files, examine the contents of files, and transform files.

In this case, I preloaded into my c:\ftp\some\ subdirectory a collection of scripts that I knew contained an "exit" line. For every file, the script substitutes the words "so long" for the exit line (not a wise change if you expect to actually execute any of the scripts in the subdirectory).

If you were writing your own multifile substitution script, you might want to change a defunct Web address wherever it appears in any file, or you might want to change a common spelling error in many files at once.

Programming languages typically provide a variety of file operations, including file tests (e.g., to determine whether a file exists or whether a directory file is a text file or a binary file), and file stats (descriptive information on the file such as file size, file creation date, or file modification date).

1.5 Counting the Words in a File

It is easy to write a short script that counts the words in a file, but it is difficult to do the job to everyone's liking. Depending on the type of text, and the intended use of the word count, the criteria for counting a word may change. For example, should numbers be counted as words? Should a Web address be counted as a word? How about e-mail addresses? Do you want to count single characters as words? Maybe you would want to include "a", "A", and "I" as words, but not "z" and "h". Or, you may want to count "A" as a word when it appears within a sentence, but not when it begins an alphabetically organized list, as in "A. Chapter 1". Because there are many way to count words, you cannot always use the word counters commonly provided in commercial word processors. There will be occasions when you will want to write your own script that counts words just as you prefer. Here is a minimalist word counting script. You can modify the script to serve your own specific needs.

1.5.1 Script Algorithm

1. For word counting exercises, we will use the OMIM® file, a well-written, public domain text corpus, described in detail in Chapter 8. The OMIM file (which exceeds 100 MB in length) is available for download by anonymous ftp from:

   ftp://ftp.ncbi.nih.gov and subdirectory: /repository/omim/omim.txt.Z

   Gunzip is a popular and open source decompression utility. If you don't have decompression software that can gunzip a gzipped file, the utility can be downloaded from http://www.gzip.org/. Gunzip the omim.txt.Z file and rename the file, for use with this script, "OMIM".

   You do not need to use the OMIM file. Feel free to substitute any plain-text file you prefer, changing the file name within the script, of course.
2. Parse through the file, line by line.
3. For each line, split the line wherever a sequence of one or more spaces is encountered, and put the resulting line fragments into an array. This has the effect of producing an array of the individual words in the line.

4. Reduce the size of the array by eliminating array items that are empty. This is necessary because splits on spaces can produce empty list items if a space precedes or ends the line.
5. Determine the size of the array. In this instance, the size of the array equals the number of words in the line.
6. Add the number of words on the current line to the running total of words in the file.
7. When you reach the end of the file, print the total number of words counted, and exit the script.

Perl Script

```
#!/usr/local/bin/perl
open (TEXT, "c\:\\big\\omim")||die"cannot";
$textvar = " ";
while ($textvar ne "")
  {
  $textvar = <TEXT>;
  @line_array = split(/[ \n]+/,$textvar);
  @reduced_array = grep($_ ne "",@line_array);
  $total = $total + scalar(@reduced_array);
  }
print $total;
exit;
```

Python Script

```
#!/usr/local/bin/python
import re
import string
total = 0
line_list = []
line_reduced = []
in_text = open('C:\\big\\omim', "r")
for line in in_text:
  line_list = re.split(r'[ \n]+',line)
  line_reduced = [var for var in line_list if var != '']
  total = total + len(line_reduced)
print total
exit
```

Ruby Script

```
#!/usr/local/bin/ruby
intext = File.open("c\:\\big\\omim", "r")
total = 0
intext.each_line(){|line| total = total + line.split.size}
puts total
exit
```

1.5.2 Analysis

The script produces the word count, for the OMIM file, currently over 20 million words, in under a minute.

1.6 Making a Word List with Occurrence Tally

Sometimes, you need to have a listing of all the different words in a document, and the number of occurrences of each word. A word frequency list tells you a lot about a document. Simple books tend to have a limited number of different words (a few thousand). Books written for advanced readers tend to have a large number of different words (20,000 or more). Word frequency lists can characterize the subject matter of a document, and can sometimes identify the author. Scanning down a list of unique words is also an excellent way to find misspellings. Misspellings can also be found by comparing the word list from your document with a word list of properly spelled words (i.e., a dictionary list). Words that are on the word list for your document, but absent from a dictionary list, are often misspelled words. In some cases, word list entries that are absent from dictionary lists are abbreviations of proper words, or names, or words from a highly specialized subject domain.

1.6.1 Script Algorithm

1. Open a text file. In this case, we will use the OMIM file.
2. Pass the entire contents of the file into a string variable. This requires computer memory that can absorb the entire file in active memory. (*Note*: If your computer cannot manage this step, you can use a smaller input file, or you can break the file into sections or lines.)
3. Match the entire text against the general pattern of a word. In this case, the general pattern of a word consists of a word-break pattern, followed by 3 to 15 letters. We make the somewhat arbitrary decision that strings less than three characters in length are either abbreviations, or they are high-frequency words of no particular interest (e.g., *of, if, in, or*). Srings with a length exceeding 15 characters are likely to be nonword letter sequences (e.g., a gene or protein sequence). The match is repeated sequentially for the entire text.
4. At each match, extract each string that matches the pattern (i.e., each word), and assign it as a key in a dictionary variable. Increment the value of the key by one. This keeps a frequency tally of each encountered word.
5. After the entire file is parsed, word by word, you are left with a dictionary wherein the keys are the different words in the text, and the key values are the frequency of occurrence of each of the different words in the text.
6. Sort the keys in the dictionary object alphabetically and print out every key–value pair to an external file.

Perl Script

```perl
#!/usr/local/bin/perl
open (OUT, ">omimword.txt");
open (TEXT, "c\:\\big\\omim");
undef($/);
$textvar = (<TEXT>);
while ($textvar =~ /(\b[a-z]{3,15}\b)/ig)
  {
  $freq{$1}++;
  }
foreach $word (sort(keys(%freq)))
  {
  print OUT $word . " - " . $freq{$word} . "\n";
  }
exit;
```

Python Script

```python
#!/usr/local/bin/python
import re
import string
word_list = []
freq_list = []
freq = {}
in_text = open('C:\\big\\omim', "r")
in_text_string = in_text.read()
out_text = open("omimword.txt", "w")
in_text_string = string.lower(in_text_string)
word_list = re.findall(r'(\b[a-z]{3,15}\b)', in_text_string)
for item in word_list:
  count = freq.get(item,0)
  freq[item] = count + 1
freq_list = freq.keys()
sort_list = sorted(freq_list)
for i in sort_list:
  print>>out_text, i, freq[i]
exit
```

Ruby Script

```ruby
#!/usr/local/bin/ruby
freq = Hash.new(0)
file_out = File.open("omimword.txt","w")
file1 = File.open("c\:\\big\\omim")
file1.read.downcase.scan(/\b[a-z]{3,15}\b/){|word| freq[word] = freq[word]+1}
freq.keys.sort.each {|k| file_out.print k, " - ", freq[k], "\n"}
exit
```

1.6.2 Analysis

We loaded the entire text of OMIM, a text file exceeding 135 MB in length, and containing about 20 million words, into a variable. The script executed in about 20 seconds, creating an output file containing about 168,000 different words, and the number of times each word occurred within the text. Here is a short sampling of the output file:

kidney—7449
kidneys—928
kido—33
kidoguchi—1
kidokoro—1
kidon—2
kidou—1
kidouchi—4
kidowaki—1
kidron—2
kidson—16

Most languages contain between 20,000 and 60,000 words. Comprehensive dictionaries, that contain many more than 60,000 words, include all of the variant forms for a single word (e.g., *soft, softer, softest, soften, softens, softening, softener*), orthographic variants (*advertise, advertize*), obsolete or denigrated variants (*publicly* and *publically*), technical words, slang, or proper names.

When we examine the OMIM output list, we see that most of the so-called words are names of people, or misspellings. If you have a 135 MB text, and a word occurs fewer than three times, it is unlikely to be a valid word. When we find a very high-frequency word, such as *with*, which occurs 212,312 times in OMIM, it is probably a low-information-content word used to connect other words. When we find a middle-frequency word, such as *kidney* (7,449 occurrences), it is almost certainly a high-information-content word relevant to the document's knowledge domain.

1.7 Using Printf Formatting Style

Printf, like regex, is another programming convention that transcends individual programming languages. Many different languages support the same printf syntax. The purpose of printf is to provide a simple way of specifying the arrangement of data printed to an output line. Printf produces output in neat columns. If you have a word, followed by three numbers, and two of the three numbers have a decimal point followed by two digits, and one of the numbers is an integer that should be left-padded with zeros to produce an integer length of 8, and you want the word and numbers in a particular order, separated by a specific number of spaces, you will want to use printf. There are numerous Web resources that will help you compose elegant printf statements.

1.7.1 Script Algorithm

1. We will use the MeSH (Medical Subject Headings) file for this exercise. In Chapter 5, we will be discussing the MeSH nomenclature, at some length. At this point, all we need to know about MeSH is that it is freely available as a large text file, from the following site:

   http://www.nlm.nih.gov/mesh/filelist.html

   Download the d2009.bin file (referred to as the ASCII MeSH download file). This plain-text file is about 28 MB in length and contains over 25,000 MeSH records. The record format of the d2009.bin file is described in the appendix.
      You can substitute any plain-text file you prefer.
      I have downloaded the file into my computer's c:\big subdirectory, and the script reflects this location.
2. Open the d2009.bin file for reading.
3. Create and open and output file, "meshword.txt".
4. Put the entire contents of the MeSH file (about 28 MB) into a variable.
5. Put each encountered word string (defined as a sequence of 3 to 15 lowercase letters) into a dictionary object key and increment its value by 1.
6. After the entire file is parsed, sort the keys, and print each key, padded to 20 spaces, followed by its value (the number of occurrences of the key in the MeSH file) as a padded six-digit integer, to the output file.

Perl Script

```
#!/usr/local/bin/perl
open (TEXT, "c\:\\big\\d2009.bin");
open (OUT, ">meshword.txt");
undef($/);
$textvar = (<TEXT>);
while ($textvar =~ /(\b[a-z]{3,15}\b)/g)
  {
  $freq{$1}++;
  }
foreach $word (sort(keys(%freq)))
  {
  printf OUT ("%-20.20s %8.06d\n", $word, $freq{$word});
  }
exit;
```

Python Script

Strictly speaking, Python has no printf function. It uses the % operator instead, but it serves the same purpose and uses an equivalent syntax.

```python
#!/usr/local/bin/python
import re
import string
word_list = []
freq_list = []
freq = {}
in_text = open('C:\\big\\d2009.bin', "r")
in_text_string = in_text.read()
out_text = open("meshword.txt", "w")
in_text_string = string.lower(in_text_string)
word_list = re.findall(r' (\b[a-z]{3,15}\b)', in_text_string)
for item in word_list:
  count = freq.get(item,0)
  freq[item] = count + 1
freq_list = freq.keys()
sort_list = sorted(freq_list)
for i in sort_list:
  print>>out_text, "%-20.20s %8.06d" % (i, freq[i])
exit
```

Ruby Script

```ruby
#!/usr/local/bin/ruby
freq = Hash.new(0)
file1 = File.open("c\:\\big\\2009.bin")
file1.read.downcase.scan(/[a-z]+/){|word| freq[word] = freq[word]+1}
freq.keys.sort.each {|k| printf "%-20.20s %8.06d\n", k, freq[k]}
exit
```

1.7.2 Analysis

Here is a partial output, showing the tail-end of the output file, listing each occurring word and the number of occurrences in the file:

```
zygosaccharomyces    000001
zygote               000023
zygotene             000006
zygotes              000008
zygout               000001
zyklon               000001
zyloprim             000001
zyloric              000001
zyma                 000005
zymafluor            000001
```

Exercises

1. Using Perl, Python, or Ruby, write a word-counting script that finds and joins words that are split by hyphens at the end of lines, and counts the connected string, as a single word.

2. Using Perl, Python, or Ruby, write a word-counting script that counts all the sentences from a file, using a period followed by one or two spaces, followed by an uppercase letter, as the sentence delimiting pattern.

3. Using Perl, Python, or Ruby, write a word-counting script that displays the first 200 and the last 200 bytes in a file of any size.

4. Using Perl, Python, or Ruby, write a script that extracts, from a file of any size, every five-word phrase that begins or contains the word *carcinoma* or *carcinomas*.

5. The word-frequency script sorts words alphabetically, with each word trailed by its frequency in OMIM. Using Perl, Python, or Ruby, rewrite the script to list the words by their decreasing frequency of occurrence.

2

UTILITY SCRIPTS

Utilities are small programs that perform a specific task, very efficiently. Most programmers treasure their personal collections of utility programs, which may include software for file compression, file searching, directory searching, file comparisons, image conversions, etc. In this chapter, we will provide some general utility scripts that will be applied in later chapters.

2.1 Random Numbers

Random numbers are used extensively in Monte Carlo simulations of biological events. The simulations are also used in statistics (e.g., calculating normal distributions), and can even provide simple computational approaches to formal mathematical problems that would otherwise require advanced numerical methods.

2.1.1 Script Algorithm

1. Create an iterator that repeats 10 times.
2. Generate and print a random number (technically, a pseudorandom number) between zero and one.

Perl Script

```
#!/usr/local/bin/perl
for (0..9)
  {
  print rand() . "\n";
  }
exit;
```

Python Script

```
#!/usr/local/bin/python
import random
for iterations in range(10):
  print random.uniform(0,1)
exit
```

Ruby Script

```
#!/usr/local/bin/ruby
(1..10).each{puts rand()}
exit
```

2.1.2 Analysis

Here is a sample output, listing 10 random numbers in the range 0 to 1:

0.594530508550135
0.289645594799927
0.393738321195123
0.648691742041396
0.215592023796071
0.663453594144743
0.427212189295081
0.730280586218356
0.768547788018729
0.906096189758145

Had we chosen, we could have rendered an integer output by multiplying each random number by 10 and rounding up or down to the closest integer.

2.2 Converting Non-ASCII to Base64 ASCII

Almost every computer user has made the mistake of trying to view a non-ASCII file (such as a binary image, or a word-processed file stored in a proprietary format) in a plain-text viewer. You will see a funny-looking page, containing happy faces and hearts, and you might even hear a few shrill beeps. Sometimes, the file, and your application will abruptly close. When a text viewer application opens a file, it converts every sequential octet of 0s and 1s (of which there are 258 possibilities) into one of the 258 standard ASCII symbols. When the file is a plain-text file, the ASCII symbols correspond to the standard keyboard symbols that we use to express written language. The standard keyboard keys account for about 64 of the 258 ASCII symbols. When the plain-text file is a non-text binary file, the octets of 0s and 1s may correspond to some of the reserved high-ASCII and low-ASCII symbols that do not appear on your keyboard.

ASCII text files, such as plain-text files, HTML files, and XML files, must only contain the standard keyboard symbols. If you want to convey a binary file (such as an image) within a text file, you must first convert the binary data to the subset of ASCII that corresponds to keyboard symbols. The converted sequence is called BASE64 encoded data. Once converted to BASE64, the text can be decoded back to binary data by reversing the algorithm. Perl, Python, and Ruby contain standard modules that will convert any file into BASE64. We will be using the BASE64 modules when we start working with image data conveyed in XML files.

2.2.1 Script Algorithm

1. Call the base64 external module into your script.
2. Read a sample file into a string variable. (In this case, the script requires you to provide a file named "z.text".)
3. Pass the string variable to the base64 encoding method provided by the module.
4. Print the base64 encoded string.
5. Pass the base64 encoded string to the decode method provided by the module.
6. Print the decoded string.

Perl Script

```perl
#!/usr/local/bin/perl
use MIME::Base64;
open (TEXT,"z.txt");
binmode TEXT;
$/ = undef;
$string = <TEXT>;
close TEXT;
$encoded = encode_base64($string);
print $encoded;
$decoded = decode_base64($encoded);
print "\n\n$decoded";
exit;
```

Python Script

```python
#!/usr/local/bin/python
import base64
sample_file = open ("z.txt", "rb")
string = sample_file.read()
sample_file.close()
print base64.encodestring(string)
print base64.decodestring(base64.encodestring(string))
exit
```

Ruby Script

```ruby
#!/usr/local/bin/ruby
require 'base64'
e_file = File.open("z.text").binmode
e_file_string = e_file.read
b64 = Base64.encode64(e_file_string)
puts b64.slice(0,300)
```

```
regular = Base64.decode64(b64)
puts regular.slice (0,300)
exit
```

2.2.2 Analysis

Here is an example of a string encoded into Base64:

This is the original string … The secret of life
This is the encoded text … VGhlIHNlY3JldCBvZiBsaWZl
This is the decoded text … The secret of life

When we use Base64, we produce output files that are larger than the original (binary) files.

2.3 Creating a Universally Unique Identifier

In Chapter 18, we will learn about Research Description Framework (RDF), a semantic device that binds metadata (data that describes data) and data to a unique information object. For RDF to have any value, information objects must be assigned a unique identifier that multiple users can apply (to the same object) across network domains. There are several strategies for applying unique identifiers to objects, one of which is the Universally Unique Identifier (UUID), also known as the Globally Unique Identifier (GUID).

The UUID is an algorithm for creating unique strings of uniform format composed of name and time information, and distributed without a central registration process.

A typical UUID may look like this:

4c108407-0570-4afb-9463-2831bcc6e4a4

The UUID can be assigned to an information object through an RDF statement and can be used to collect information related to an object, or any of the objects from the same class as the unique object or from any of the classes related to the class of the unique object (more on this in Chapter 18).

Unique identifiers are described at length in my book *Biomedical Informatics* (Jones & Bartlett Publishers, 2007, pp. 78–85).

Perl, Python, and Ruby all have modules that will generate UUIDs.

2.3.1 Script Algorithm

1. Call the external module that creates UUID strings.
2. Create a new UUID object.
3. Print the UUID string.

Perl Script

The UUID::Tiny module is available from the Perl Packet Manager (ppm).

```
#!/usr/local/bin/perl
use UUID::Tiny;
print create_UUID_as_string();
exit;
```

Python Script

The UUID module is included in the standard python distribution and can be called directly from the script.

```
#!/usr/local/bin/python
import uuid
print uuid.uuid4()
exit
```

Ruby Script

The GUID module is available as a gem and can be installed with the following command:

```
gem install guid
```

```
#!/usr/local/bin/ruby
require 'guid'
puts Guid.new
exit
```

2.3.2 Analysis

The algorithms for creating UUIDs, and all of the standard versions of the algorithm, are described in a publicly available Request for Comments file:

http://www.ietf.org/rfc/rfc4122.txt

2.4 Splitting Text into Sentences

Many text parsing algorithms proceed sentence by sentence, not line by line. This is important in machine translation and natural language exercises that use grammar rules to extract concepts whose parts are scattered through the sentence.

Not infrequently, the information specialist begins a script by teasing out the sentences from a narrative text.

2.4.1 Script Algorithm

1. Start with a variable containing text.
2. Split the text wherever there is an occurrence of a period (or other sentence delimiter, such as a question mark or quotation mark), followed by one or more spaces, followed by an uppercase letter.
3. Put the resulting sequences (which should consist largely of sentences) into an array.

Perl Script

```
#!/usr/local/bin/perl
$all_text = "I am here. You are here. We are all here.";
$all_text =~ s/([^A-Z]+\.[ ]{1,2})([A-Z])/$1\n$2/g;
print $all_text;
exit;
```

Python Script

```
#!/usr/local/bin/python
import re
all_text = "I am here. You are here. We are all here.";
sentence_list = re.split(r'[\.\!\?] +(?=[A-Z])', all_text)
print '\n'.join(sentence_list)
exit
```

Ruby Script

```
#!/usr/local/bin/ruby
all_text = "I am here. You are here. We are all here.";
all_text.split(/[\.\!\?] +(?=[A-Z])/).each {|phrase| puts phrase}
exit
```

2.4.2 Analysis

The input is:

"I am here. You are here. We are all here."

The output is:

I am here
You are here
We are all here.

Notice that only the last sentence is terminated by a period. This is because the last sentence does not match the regex pattern. The period of the last sentence is not followed by one or more spaces and an uppercase letter. The last sentence is included in the output only because it was split from the prior match. There are many ways by which we could have corrected for this particular limitation, but sometimes a programmer needs to decide when the performance of a less-than-perfect script is sufficient for his intended purposes. In the case of our sentence parser, some sequences that are not sentences would be split off by the regex pattern. One example is a number, followed by a period, followed by spaces and an uppercase letter. You may encounter such a sequence in a list. In addition, sentences are often separated by whitespace characters other than the space character (such as a newline character or a tab character). Our sentence parser does not account for these types of legitimate sentence delimiters. The point is that programmers must be ready to redesign their scripts to account for the kinds of textual material they are likely to encounter. Expanding a regex pattern or adding additional pattern filters will slow the script and will almost certainly have unintended consequences (requiring more programming). We will be using this simple sentence parsing pattern in later scripts, as we parse through large public domain text files. We can always incrementally modify the sentence parsing patterns when we inspect the output from our scripts.

2.5 One-Way Hash on a Name

A one-way hash is an algorithm that transforms a string into another string in such a way that the original string cannot be calculated by operations on the hash value (hence the term "one-way" hash). Examples of public domain one-way hash algorithms are MD5 and the Secure Hash Algorithm (SHA). These differ from encryption protocols that produce an output that can be decrypted by a second computation on the encrypted string.

The resultant one-way hash values for text strings consist of near-random strings of characters, and the length of the strings (e.g., the strength of the one-way hash) can be made arbitrarily long. Therefore, name spaces for one-way hashes can be so large that the chance of hash collisions (two different names or identifiers hashing to the same value) is negligible. For the fussy among us, protocols can be implemented guaranteeing a data set free of hash collisions, but such protocols may place restrictions upon the design of the data set.

In theory, one-way hashes can be used to anonymize patient records while still permitting researchers to accrue data over time to a specific patient's record. Names of patients and other identifiers are replaced by their one-way hash values. If a patient returns to the hospital and has an additional procedure performed, the record identifier, when hashed, will produce the same hash value held by the original data set record. The investigator simply adds the data to the "anonymous" data set record containing the same one-way hash value. Since no identifier in the anonymized data set record can be used to link back to the patient, confidentiality is preserved.

Quantin and Bouzelat have standardized a protocol for coding names using SHA one-way hashes.[*]

There is no practical algorithm that can take an SHA hash and determine the name (or the social security number or the hospital identifier, or any combination of the above) that was used to produce the hash string. In France, the name-hashed files are merged with files from many different hospitals and used in epidemiologic research. They use the hash codes to link patient data across hospitals. Their methods have been registered with Service Central de la Securite des Systemes d'information (SCSSI).

Implementation of one-way hashes involve certain practical problems. Attacks on one-way hash data may take the form of hashing a list of names and looking for matching hash values in the data set. This type of attack, the so-called dictionary attack, can be countered by encrypting the hash or by hashing a secret combination of identifier elements or both or keeping the hash value private (hidden). Issues arise related to the multiple ways that a person may be identified within a hospital system (Tom Peterson on Monday, Thomas Peterson on Tuesday), resulting in inconsistent hashes on a single person. Resolving these problems is an interesting area for further research.

The text in this section is an excerpt from a public domain document (Berman J. J. Confidentiality for medical data miners. *Artificial Intelligence in Medicine* 26:25–36, 2002.)

2.5.1 Script Algorithm

1. Call the external MD5 module from your script.
2. Prompt the user to enter a name.
3. Collect the entered name, remembering to strip the newline terminator (produced by pressing the return key).
4. Pass the entered phrase to the MD5 method module.
5. Print the returned one-way hash value, in hexadecimal format.

Perl Script

```
#!/usr/local/bin/perl
use Digest::MD5 qw(md5 md5_hex md5_base64);
print "What is your full name?\n";
$phrase = <STDIN>;
$phrase =~ s/\n//;
print md5_hex($phrase);
exit;
```

---

[*] Quantin C., Bouzelat H., et al. Automatic record hash coding and linkage for epidemiological follow-up data confidentiality. Meth Inf Med 37:271–277, 1998.

Two sample script executions:

```
What is your full name?
jules j berman
Output:
f1f78b23dea43a15ffd73ab7d5731022

What is your full name?
Jules J Berman
Output:
c9941f26601fe1d2183ba05dd2a199ee
```

Python Script

```
#!/usr/local/bin/python
import sys, string, md5
print "What is your full name?"
line = sys.stdin.readline()
line = line.rstrip()
md5_object = md5.new()
md5_object.update(line)
print md5_object.hexdigest()
exit
```

Two sample script executions:

```
What is your full name?
jules j berman
Output:
f1f78b23dea43a15ffd73ab7d5731022

What is your full name?
Jules J Berman
Output:
c9941f26601fe1d2183ba05dd2a199ee
```

Ruby Script

```
#!/usr/local/bin/ruby
require 'digest/md5'
puts "What is your full name?"
phrase = gets.chomp
hexstring = Digest::MD5.hexdigest(phrase)
puts hexstring
exit
```

Two sample script executions:

```
What is your full name?
jules j berman
Output:
f1f78b23dea43a15ffd73ab7d5731022
```

```
What is your full name?
Jules J Berman
Output:
c9941f26601fe1d2183ba05dd2a199ee
```

2.5.2 Analysis

There are several available one-way hash algorithms. MD5 is available as a standard module for many different programming languages, but the SHA algorithm is also available. Notice that the output is case-sensitive. The hash value for "jules j berman" is completely different from the hash value of "Jules J Berman." Those who wish to substitute a hash value for a name must be careful to use a consistent format for each name.

2.6 One-Way Hash on a File

All values produced by the one-way hash algorithm are fixed-length. The one-way hash value for a 10 megabyte (MB) file will have the same length as a one-way hash value for a patient's name. A change of a single character in a file will result in a completely different one-way hash value for the file.

By sending a one-way hash value for a file, along with the file itself, you can, with a high degree of confidence, authenticate your file. When others receive your file, along with its MD5 hash that you created, they can recompute the MD5 hash on the file and compare the output with the MD5 hash that you sent. If the two hash numbers are identical, then they can be fairly certain that the file was not altered from the original file (for which the original MD5 value was computed).

Of course, there is always the possibility that the file and its one-way hash were intercepted en route, and that both file and hash were altered. In this case, comparing hashes will not help. A variety of transmission protocols have been developed to thwart simple interception attacks.

2.6.1 Script Algorithm

1. Import an external standard module that computes the MD5 one-way hash.
2. Read the contents of the file into a string. In this case, we use the file, "us.gif," but any file would suffice.
3. Call the module function to create the MD5 digest value on the contents of the file.
4. Print out the digest value in hex format.

Perl Script

```
#!/usr/local/bin/perl
use Digest::MD5 qw(md5 md5_hex md5_base64);
```

```
open (TEXT,"us.gif");
binmode TEXT;
$/ = undef;
$line = <TEXT>;
print md5_hex($line);
exit;
```

Python Script

```
#!/usr/local/bin/python
import md5
import string
md5_object = md5.new()
sample_file = open ("us.gif", "rb")
string = sample_file.read()
sample_file.close()
md5_object.update(string)
md5_string = md5_object.digest()
print ''.join([ "%02X " % ord( x ) for x in md5_string]).strip()
exit
```

Ruby Script

```
#!/usr/local/bin/ruby
require 'digest/md5'
file_contents = File.new("us.gif").binmode
hexstring = Digest::MD5.hexdigest(file_contents.read)
puts hexstring
exit
```

2.6.2 *Analysis*

The script's output is

    39842f5ed1516d7c541155fd2b093b36

The alphanumeric sequence is the MD5 message digest of the us.gif image file. Changing a single byte in the original file, and repeating the MD5 digest operation will yield an entirely different digest value.

2.7 A Prime Number Generator

A prime number, by definition, cannot be the product of two integers. If a number is prime, then there will be no smaller number that will divide into the number without producing a remainder. To determine if a number is prime, we can test each smaller number, to see if it divides into the number without leaving a remainder. If not, then the number is a prime.

We can use a little trick to shorten the process, by stopping the iterations when we have examined every smaller number in ascending order up to the square root of the number. If there were an integer larger than the square root of the number that could be multiplied by another integer to give the number, then the other integer would need to be smaller than the square root of the number (otherwise, the two integers would produce a product larger than the number). But we have already tested all of the numbers smaller than the square root of the number, and they all yielded a non-zero remainder. So we do not need to test the integers greater than the square root of the number.

Here is how you can generate a very long list of prime numbers with just a few lines of code.

2.7.1 Script Algorithm

1. Create a loop for all of the integers up to an arbitrary maximum (1,000 in this case).
2. If an integer is prime, then there will be no smaller integer that will divide into the number with a remainder of 0. For each number in the outer loop, loop through the numbers smaller than the integer (a nested loop) to see if any of the smaller numbers divides into the larger number with anything other than a zero remainder. If not, then the larger number must be a prime.
3. As we loop through the smaller numbers (the nested loop numbers), we can save some computational time by stopping when we reach the square root of the larger number. If there were an integer larger than the square root of the number, which could be multiplied by another integer to give the number, then the other integer would need to be smaller than the square root of the number (otherwise, the two integers would produce a product larger than the number). But we've already tested all of the numbers smaller than the square root of the number, and they all yielded a nonzero remainder. So we don't need to test the integers greater than the square root of the number.

Perl Script

```
#!/usr/local/bin/perl
print "2,3,";
for($i=4;$i<10000;$i++)
    {
    for $thing (2 .. int(sqrt($i)))
        {
        $state = 1;
        if ($i % $thing == 0)
            {
            $state = 0;
            last;
            }
```

```
        }
    print "$i\," unless ($state == 0);
    }
exit;
```

Python Script

```
#!/usr/local/bin/python
import math
print "2,3,"
state = 1
for i in range(4, 10000):
    upper = math.sqrt(i)
    upper = int(upper)
    for thing in range(2, upper):
        state = 1
        if (i % thing == 0):
            state = 0
            break
    if (state == 1):
        print i,
exit
```

Ruby Script

```
#!/usr/local/bin/ruby
state = Numeric.new
print "2,3,"
(4..10000).each do
    |i|
    (2..(Math.sqrt(i).ceil)).each do
        |thing|
        state = 1
        if (i.divmod(thing)[1] == 0)
            state = 0
            break
        end
    end
    print "#{i}\," unless (state == 0)
end
exit
```

Ruby has an even simpler way to produce primes, using the built-in Prime Class.

```
#!/usr/local/bin/ruby
require 'Mathn'
generator = Prime.new
count = 0
generator.each{|i| puts i; count = count + 1; break if count > 100}
exit
```

2.7.2 Analysis

Every biomedical scientist who uses medical records and other confidential data can benefit by understanding the role of prime numbers. Almost every cryptographic method relies on methods that produce large prime numbers, which, when multiplied together, produce a number that cannot be factored by a quick computation.

Here is the partial output of our method for producing prime numbers:

2,3,5,7,11,13,17,19,23,29,31,37,41,43,47,53,59,61,67,71,73,79,83,89,97,101,1 03,107,109,113,127,131,137,139,149,151,157,163,167,173,179,181,191,193,19 7,199,211,223,227,229,233,239,241,251,257,263,269,271,277,281,283,293,307 ,311,313,317,331,337,347,349,353,359,367,373,379,383,389,397,401,409,419,42 1,431,433,439,443,449,457,461,463,467,479,487,491,499,503,509,521,523,541,5 47,557,563,569,571,577,587,593,599,601,607,613,617,619,631,641,643,647,653, 659,661,673,677,683,691,701,709

Exercises

1. In the script from Section 2.2, we converted a file to Base64. Modify the script to query the user for a string, then write (to the monitor) the Base64 encoded string.
2. Sentences can end in characters other than a period (such as "!" or "?", or quotation marks). Rewrite the sentence parser to include these sentence terminators.
3. The biggest weakness of every one-way hash algorithm is the dictionary attack. Suppose you have a one-way hash sequence, composed of seemingly random letters and numbers. You would like to know the word, phrase, or file that, when hashed, produced the sequence. You know that there is no way to produce the original text by examining the sequence. However, you also know that there are a limited number of sources for the sequence. If the sequence represents a one-way hash on an English word, you can perform the one-way hash on every word in the English language (from a word list). When you come to a word that produces a hash sequence that is identical to the sequence you are trying to analyze, then you know that the word must be source of the sequence. Likewise, if you know that the one-way hash sequence represents the name of a person, you can start with a list of names (for example, the names of a phone book) and determine the one-way hash for each of the names in the list, stopping when you find a match to the hash-sequence at-hand.
   Here are five one-way hash sequences produced by five English words:
   a4704fd35f0308287f2937ba3eccf5fe
   f85b785512fe9685dde7fda470fe2b9f
   c30635cc93c51c6f6731806dbd149a51
   a7c291d9a3df61237f8415d5f0149ac7
   a58a246f88b5442d91c01fadc4bb7831
   5fed3411faf832174ef1f040028b2c21
   Using Perl, Python, or Ruby, write a script that discovers the original five words that yielded the five hash sequences.

Hint: Parse through OMIM, selecting each different word, and producing three MD5 hash sequences for each word (the lowercase word,
the word with the first letter uppercase and the subsequent letters
lowercase, and the word with all letters uppercase). For each word
encountered, check to see if there is a match to any of the five provided one-way hash sequences. Repeat until the unique words in the
OMIM file are exhausted.

4. What well-known and universal constant is approximated by the following
script (in Perl, Python or Ruby)? Explain how the scripts works.

```perl
#!/usr/local/bin/perl
for (1..10000000)
  {
  $x = rand();
  $y = rand();
  $r = sqrt(($x*$x) + ($y*$y));
  if ($r < 1)
    {
    $totr = $totr + 1;
    }
  $totsq = $totsq + 1;
  }
print eval(4 * ($totr / $totsq));
exit;
```

```python
#!/usr/bin/python
import random
from math import sqrt
totr = 0
totsq = 0
for iterations in range(1000000):
  x= random.uniform(0,1)
  y= random.uniform(0,1)
  r= sqrt((x*x) + (y*y))
  if r < 1:
    totr = totr + 1
  totsq = totsq + 1
print float(totr)*4.0/float(totsq)
exit
```

```ruby
#!/usr/local/bin/ruby
x = y = totr = totsq = 0.0
(1..100000).each do
  x = rand()
  y = rand()
  r = Math.sqrt((x*x) + (y*y))
  totr = totr + 1 if r < 1
  totsq = totsq + 1
end
puts (totr *4.0 / totsq)
exit
```

5. Because an entire file can be represented as a one-way hash value, your friend
Bob was inspired with an idea. Bob will compute the file's one-way hash value

and then insert the value into the top line of the file. He sends the file to you. You perform a one-way hash operation on the received file. If your one-way hash exactly matches the one-way hash inserted into the top line of the file, then you know that the file has not been altered during transit. This is actually a terrible idea and cannot possibly succeed. Why not?

3

VIEWING AND MODIFYING IMAGES

Everyone who deals with data will eventually need a simple way of representing their data in images. There are many different image formats that are currently available. Here are a few:

JPEG—Joint Photographic Experts Group, a compressed image format, particular suited to photographs, commonly used by digital cameras, and appearing on many Web pages.

PNG—Portable Network Graphics, an image format created as a patent-free alternative to GIF. GIF contains a patented compression algorithm (LZW). Though the patent used by GIF has now expired, PNG still serves as a popular GIF-like format, used extensively on Web pages.

TIFF—Tagged Image File Format, an image and line-drawing file format that is widely supported by page layout, character recognition, and publishing applications.

GIF—Graphics Interchange Format, an image format, first introduced by Compuserve, particularly well-suited for schematic and line-drawing images, that is used extensively on Web pages.

DICOM—Digital Imaging and Communications in Medicine, a medical imaging format used primarily in radiology services.

The most popular image format on the Web is JPEG. This is the format used in most digital cameras, and there are billions of JPEG images distributed through the Internet.

Here is how you can display any JPEG image on your monitor. We will be using this technique in later chapters, when we create charts, graphs, and mashups to visualize our work.

3.1 Viewing a JPEG Image

Every programming language should have a method of creating a "window" on your monitor, in which you can view images, text, and media. Perl, Python, and Ruby all use Tk, an open source utility that provides functional widgets (buttons, menus, canvases, text boxes, frames, labels, and so on). By calling the Tk module from Perl, Python, or Ruby, you can create applications with a sophisticated graphic user interface (GUI). With Tk, you can write scripts that display images.

3.1.1 Script Algorithm

1. Call the Tk module.
2. Create a window widget.
3. Create an image object, supplying the filename of an image to view. We could have used any image in a wide variety of image formats. In this case, we will use 3320.jpg. The image file can be downloaded at:

   http://www.julesberman.info/book/3320.jpg

4. Pack the image into the widget.
5. Begin the Tk operational loop.

Perl Script

```
#!/usr/local/bin/perl
use Image::Magick;
my $im_fna = Image::Magick->new;
$im_fna -> ReadImage("c\:\\ftp\\3320.jpg");
$im_fna -> write ("gif:c\:\\ftp\\3320.gif");
$im_fna -> resize ("0.4");
use Tk;
$mw = MainWindow->new();
$image = $mw->Photo(-file => "c\:\\ftp\\3320.gif");
$mw->Label(-image=>$image)->pack;
MainLoop;
exit;
```

Python Script

```
#!/usr/local/bin/python
import Tkinter
import Image, ImageTk
im_fna = Image.open('c:/ftp/3320.jpg')
im_fna.save('c:/ftp/3320.gif')
im_fna = im_fna.resize((400,400))
root = Tkinter.Tk()
tkim_fna = ImageTk.PhotoImage(im_fna)
Tkinter.Label(root, image=tkim_fna).pack()
root.mainloop()
exit
```

Python permits an even simpler method for viewing images, using PIL (Python Image Library). PIL is freely available from

   http://www.pythonware.com/products/pil/

We can easily display an image by calling the show() method provided in PIL (the Python Image library).

```
#!/usr/local/bin/python
import Image
im=Image.open("c:/ftp/neo1.gif")
im.show()
exit
```

The show() method makes a surreptitious operating system call to your default image viewing application. This means that if your operating system has no default image viewer, or if the operator system cannot call an image viewer, or if the image viewer, for any reason, cannot respond properly to a call to display an image, this script will not work.

Python's TKinter will provide better control over the images you display than will PIL's show() method.

Ruby Script

You will need to install RMagick, Ruby's open source interface to the ImageMagick open source library. Instructions for obtaining RMagick and ImageMagick are available in the appendix of this book.

```
#!/usr/local/bin/ruby
require 'RMagick'
include Magick
im_fna = ImageList.new("c:/ftp/3320.jpg").resize!(0.4)
im_fna.write("c:/ftp/3320.gif")
require 'tk'
root = TkRoot.new {title "Ruby View"}
TkButton.new(root) do
  image TkPhotoImage.new{file "c:/ftp/3320.gif"}
  command {exit}
  pack
end
Tk.mainloop
exit
```

3.1.2 Analysis

The script produces a window displaying the image (Figure 3.1). Perl, Python, and Ruby produce the same Tk window object, but with a language-specific heading in the window bar (top).

With a few lines of code, you can view any selected image. With this basic functionality, you can build complex graphical user interfaces that display images selected from a list or multiple images.

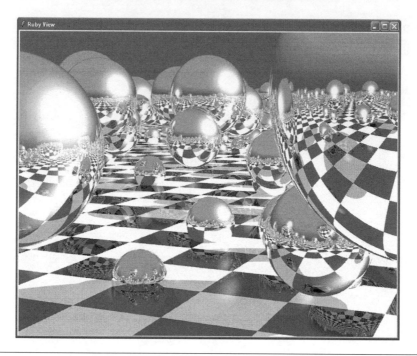

Figure 3.1   Sample image from script, displayed with Ruby's Tk interface.

3.2 Converting between Image Formats

Most people choose one image format that they use consistently for the bulk of their work. Often, this will be the format that best displays the kinds of images they create or capture in their customary projects. Those who use images containing texture, lighting variations, and many small detailed objects may prefer JPEG files. Those who create line drawings and schematics may prefer GIF or PNG. Those who combine publication-quality pages, mixing images, and text may prefer PDF. Rather than learn new tricks, it is often easiest to convert images to your preferred format. When you want to export images to colleagues who prefer another format, you will need software that reverses the process.

3.2.1 Script Algorithm

1. Call the image library into your script.
2. Create a new image object, and provide the name of an image file to the new image object.
3. Create another image object.
4. Write the second image object to whatever image formats you prefer, by assigning the preferred suffix to the filename.

Perl Script

```perl
#!/usr/local/bin/perl
use Image::Magick;
my $image = Image::Magick->new;
$image -> ReadImage("neo1.jpg");
$image -> write ("gif:neo1.gif");
$image -> write ("png:neo1.png");
$image -> write ("pdf:neo1.pdf");
exit;
```

Python Script

```python
#!/usr/local/bin/python
import Image
im = Image.open("neo1.jpg")
im.save("neo1.gif")
im.save("neo1.png")
im.save("neo1.pdf")
exit
```

Ruby Script

```ruby
#!/usr/local/bin/ruby
require 'RMagick'
include Magick
image = ImageList.new("neo1.jpg")
image_copy = image.copy
image_copy.write("neo1.png")
image_copy.write("neo1.pdf")
image_copy.write("neo1.gif")
exit
```

3.2.2 Analysis

Though it is easy to convert between different image formats, you should keep in mind that the specifications for the varying image formats are constantly changing. The version of an image format produced by your Perl, Python, or Ruby script may not be the version required by your specialized image applications. When images are converted between image formats, unexpected modifications in the image may result. It is good practice to always save the image that you start with, making your conversions on a copy of the original image. Be prepared to lose some information, particularly text annotations placed in the header of the image, when the image is converted to another format.

3.3 Batch Conversions

When you write your own image software, you can automate activities that would otherwise require repeated operations, on multiple image files, with off-the-shelf image processing software. For example, you might want to delete, add, or modify annotations for a group of images, or you might want to resize an image collection to conform to specified dimensions. When you have more than a few images, you will not want to repeat the process by hand, for each image. When you have thousands of images, stored in a variety of image formats, it will be impossible to implement global conversions, if you do not know how to batch your operations.

Here is an example of a script that converts a batch of images from color to grayscale.

3.3.1 Script Algorithm

1. Import the image module for your programming language.
2. For this example, the source images are all located in the c:\ftp\rgbfigs\ subdirectory. Every file in the subdirectory is an image file. The images are color images in .jpg, .gif, or .png formats.
3. Collect the names of all of the images in the c:\ftp\rgbfigs\subdirectory.
4. Loop through each image name in the subdirectory.
5. If the image name contains the suffix ".db", go to the next image name. This step is required because the exported image modules will insert a file with a .db extension into the image subdirectory. The .db file is not an image file and cannot be converted to a grayscale image. You will need to ignore this file.
6. For each image file in the subdirectory, create an image object.
7. Convert the image object to grayscale.
8. Write the image object to a new file, with the same name, in another sub-directory. In this script we use c:\ftp\bwfigs\ as the target subdirectory.
9. Repeat for each image.

Perl Script

```
#!/usr/local/bin/perl
use Image::Magick;
@array = glob("c:\\ftp\\rgbfigs\\*.*");
foreach $file_path (@array)
 {
 next if $file_path =~ /\.db/;
 $file_path =~ /\\([a-z0-9\_\.\-]+)$/i;
 $filename = $1;
 $image = Image::Magick->new;
 $image -> ReadImage($file_path);
 $image->Quantize(colorspace=>'gray');
```

```
$file_new = "c:\\ftp\\bwfigs\\" . $filename;
$image -> Write ($file_new);
}
exit;
```

Python Script

```
#!/usr/local/bin/python
import Image, sys, os, re
filelist = os.listdir("c:\\ftp\\rgbfigs")
os.chdir("c:\\ftp\\rgbfigs")
for file in filelist:
  if re.search('\.db', file):
    continue
  im = Image.open(file).convert("L")
  newfile = "c:\\ftp\\bwfigs\\" + file
  im.save(newfile)
exit
```

Ruby Script

```
#!/usr/local/bin/ruby
require 'RMagick'
include Magick
filelist = Dir.glob("c:/ftp/rgbfigs/*.*")
filelist.each do
 |filepathname|
 filepathname = filepathname.chomp
 filepathname =~ /c:\/ftp\/rgbfigs\//
 filename = $'
 next if filename =~ /\.db/
 newpathname = "c:/ftp/bwfigs/" + filename
 bw_image = ImageList.new(filepathname).quantize(number_colors=256,
    colorspace=Magick::GRAYColorspace, dither=true, tree_depth=0,
    measure_error=false)
 bw_image.write(newpathname)
end
exit
```

3.3.2 Analysis

The script produces grayscale versions of all the images in the c:\ftp\rgbfigs subdirectory and inserts them in the c:\ftp\bwfigs subdirectory. Conversion from color to grayscale is somewhat slow. If you have hundreds of images, the script may take longer than a minute to execute.

In this example, we chose a simple function, grayscale conversion. There are hundreds of ImageMagick functions, and we could have written a script that employs many different conversion steps on every image.

3.4 Drawing a Graph from List Data

One of the simplest and most useful ways of visualizing list data (i.e., arrays of data) is the bar graph. The task of converting data lists into bar graphs is so common that most spreadsheet applications, and some word processing applications, will build graphs from data. Nonetheless, serious informaticians should know how to build a bar graph from their own data sources. When you write your own scripts for building graphs, you have the flexibility to automate and modify the process of graph creation. You can build graphs from very large data arrays or from data extracted from multiple, diverse data sets, and you can create hundreds or thousands of graphs, virtually instantaneously, from multiple data arrays. You can write scripts that build graphs and export the graphs into Web pages. All of these efforts begin with the simple technique of converting a list of numbers into a bar graph.

3.4.1 Script Algorithm

1. Import an image library.
2. Open (or create) a blank image object. In this case, we use a blank GIF image, named empty.gif. It is a simple 500 pixel by 500 pixel image, with a white background. If you would like to use the same image for your own scripts, you can download it at:

   http://www.julesberman.info/book/empty.gif

3. Split the input data list into an array. In this case, we use the following input data list:

   "1 1 1 3 4 9 27 45 89 89 32 51 69 92 11 11 80 43"

4. Use each array item as a "height" quantity to be added to the baseline y-coordinate for the graph, and increment each x-coordinate by some predetermined number for each successive array item.
5. Use the image library's draw method to add lines to your image.
6. Write your image object to a new image file.

Perl Script

Uses Image::Magick, available as an external module from the ActiveState ppm service.

```
#!/usr/local/bin/perl
use Image::Magick;
my $image = Image::Magick->new;
$image -> ReadImage("c\:\\ftp\\metajpg\\empty.gif");
```

```
$data_string = "1 1 1 3 4 9 27 45 89 89 32 51 69 92 11 11 80 43";
@data_array = split(/ /,$data_string);
$x_coordinate = 15;
foreach $thing (@data_array)
  {
  $x_coordinate = $x_coordinate + 25;
  $y_coordinate = 300 - $thing;
  $image -> Draw (stroke => "black", width => "2", primitive => "line",
  points => "$x_coordinate,300 $x_coordinate,$y_coordinate");
  }
$image -> write ("empty1.gif");
exit;
```

Python Script

```
#!/usr/local/bin/python
import Image, ImageDraw
im = Image.open("c:/ftp/metajpg/empty.gif")
draw = ImageDraw.Draw(im)
data_string = "1 1 1 3 4 9 27 45 89 89 32 51 69 92 11 11 80 43"
data_array = data_string.split(" ")
x_coord = 20
for i in data_array:
  x_coord = x_coord + 25
  y_coord - 300 - int(i)
  draw.line((x_coord,300) + (x_coord, y_coord), width=4, fill=000)
im.save("c:/ftp/metajpg/empty1.jpg")
exit
```

Ruby Script

```
#!/usr/local/bin/ruby
require 'RMagick'
include Magick
img = Magick::ImageList.new("c:/ftp/metajpg/empty.gif")
gc = Magick::Draw.new
gc.fill_opacity(0)
gc.stroke('black').stroke_width(3)
data_string = "1 1 1 3 4 9 27 45 89 89 32 51 69 92 11 11 80 43"
data_array = data_string.split(/ /)
baseline = 30
data_array.each do
  |x|
  gc.line(baseline,(300 - x.to_i), baseline,300)
  baseline = 25 + baseline
end
gc.draw(img)
img.write("c:/ftp/empty1.gif")
exit
```

Figure 3.2   The graph produced from the input array, "1 1 1 3 4 9 27 45 89 89 32 51 69 92 11 11 80 43".

3.4.2 Analysis

The script produces an image, containing a simple graph, without designated coordinates, for the provided list data (Figure 3.2).

Image Magick supports hundreds of operations for drawing images and modifying existing images. Most of the Image Magick operations can be directly called from your own scripts. The Image Magick Web site is

http://www.imagemagick.org/script/index.php

3.5 Drawing an Image Mashup

A mashup is a computational trick that uses complex data from one or more sources and displays the data in a new context, often employing images to create a simplified representation of data or concepts. Web-based mashups use the power of the Web to draw information (news, images, etc.) from multiple sources to produce dynamic services. Simple mashups, such as the ones that we will create here, take information from one or more data set files and present the aggregated information in charts or images.

3.5.1 Script Algorithm

1. Import an image module into your script.
2. Determine the northern and southern latitudes, and the eastern and western longitudes that mark the perimeter of the United States.
3. Open the external file (loc_states.txt) that contains the map coordinates for the geographic centers of each state. The first few lines of the file are shown here, with the latitude and longitudes for Alaska, Alabama, and Arkansas.

"AK,61.3850,-152.2683"
"AL,32.7990,-86.8073"
"AR,34.9513,-92.3809"

These three lines mean the following:

Alaska Latitude 61.3850 (North) Longitude 152.2683 (West)
Alabama Latitude 32.7990 (North) Longitude 86.8073 (West)
Arkansas Latitude 34.9513 (North) Longitude 92.3809 (West)

The file was obtained from the following source:

http://www.maxmind.com/app/state_latlon

State longitudes and latitudes, obtained from the state_latlon file, and used in this script, are available at

http://www.julesberman.info/book/loc_states.txt

For this script, we deposited loc_states.txt in the c:\ftp\ subdirectory of our hard drive.

4. Create two dictionary objects. In both, the two-letter state codes are the keys. In one, the values are the latitude locations of the states. In the other, the values are the longitude locations of the states.

5. Open an image file consisting of a map of the United States. We use here the public domain image obtained from the U.S. National Oceanic and Atmospheric Administration.

http://www.nssl.noaa.gov/papers/techmemos/NWS-SR-193/images/fig7.gif

I "erased" the interior of the map, leaving a minimalist outline of the United States upon which to project the state-specific data. You can use any map, so long as you know the longitude and latitude boundaries.

In general, this method works best with maps of modest geographic content (i.e., less than or equal to the size of the United States). The reason for this is that the algorithm requires a rectangular coordinate system. For large areas of the earth, surface curvature makes this difficult. When you attempt to project latitude-longitude points onto large area maps, simple proportionate scale can produce strange results. This is not a problem for maps that cover a small surface (i.e., a few hundred miles).

The file used in this sample script (Figure 3.3) is available for download at

http://www.julesberman.info/book/us.gif

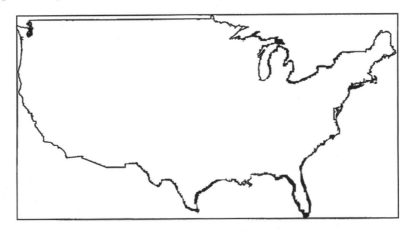

Figure 3.3   The input image, the outline of the (contiguous) United States.

6. The coordinates of the perimeter of the map are as follows:

> North = 49 degrees; #northernmost latitude of map in degrees north
> South = 25 degrees; #southernmost latitude of map in degrees north
> West = 125 degrees; #westernmost longitude of map in degrees west
> East = 66 degrees; #easternmost longitude of map in degrees west

The location of each state can be positioned to a specific point on the map by calculating the fraction of the map's north–south and east–west distances (in degrees) that is occupied by each state's latitude and longitude.

7. Determine the number of columns and rows in the map image. This gives you the width (columns) and height (rows) of the full image.

8. For each state, translate the global coordinates for each state as *x,y* coordinates on the map image.

9. Draw circles on the map, using the *x,y* coordinates for each state as the center for each circle.

10. Write the resulting image to an external image file.

Perl Script

```perl
#!/usr/local/bin/perl
use Image::Magick;
$north = 49; #Northernmost latitude of map in degrees north
$south = 25; #Southernmost latitude of map in degrees north
$west = 125; #Westernmost longitude of map in degrees west
$east = 66; #Easternmost longitude of map in degrees west
open(TEXT, "c\:\\ftp\\loc_states.txt");
$line = " ";
while ($line ne "")
    {
    $line = <TEXT>;
    $line =~ /^([A-Z]{2})\,([0-9\.]+)\,\-?([\.0-9]+) *$/;
    $state = $1;
    $latitude = $2;
    $longitude = $3;
    $lathash{$state} = $latitude;
    $lonhash{$state} = $longitude;
    }
my $img1 = Image::Magick->new;
$img1 -> ReadImage("us.gif");
$width = $img1 -> Get('columns');
$height = $img1 -> Get('rows');
print "$width $height\n";
while ((my $key, my $value) = each(%lathash))
    {
    $state = $key;
```

```
    $latitude = $value;
    $longitude = $lonhash{$key};
    $offset_y = int(((($north - $latitude) / ($north - $south)) *
$height);
    $offset_x = int(((($west - $longitude) / ($west - $east)) *
$width);
    $radius = $offset_x - 15;
    $img1 -> Draw (
      stroke => "red",
      primitive => "circle",
      points => "${offset_x}, ${offset_y}, ${radius}, ${offset_y}");
    }
$img1 -> write ("gif:us_out.gif");
exit;
```

Python Script

```python
#!/usr/local/bin/python
import sys
import Image, ImageDraw
import re
lathash = {}
lonhash = {}
north = 49
south - 25
west = 125
east = 66
infile = open ("loc_states.txt", "r")
for line in infile:
    match_tuple = re.match(r'^([A-Z]{2})\,([0-9\.]+)\,\-?([\.0-9]+)
*$',line)
    state = match_tuple.group(1)
    latitude = float(match_tuple.group(2))
    longitude = float(match_tuple.group(3))
    lathash[state] = latitude
    lonhash[state] = longitude
im = Image.open("us.jpg")
print im.mode
[width, height] = im.size
draw = ImageDraw.Draw(im)
for state, latitude in lathash.iteritems():
    longitude = lonhash[state]
    offset_y = int(((north - latitude) / (north - south)) * height)
    offset_x = int(((west - longitude) / (west - east)) * width)
    print offset_x,offset_y
    draw.ellipse((offset_x, offset_y, (offset_x + 10), (offset_y +
10)), outline=0xff0000, fill=0x0000ff)
im.save("us_out.jpg")
exit
```

Ruby Script

```
#!/usr/local/bin/ruby
require 'RMagick'
north = 49.to_f #degrees latitude
south = 25.to_f #degrees latitude
west = 125.to_f #degrees longitude
east = 66.to_f #degrees longitude
#corresponds to the us continental extremities
text = File.open("c\:\\ftp\\loc_states.txt", "r")
lathash = Hash.new
lonhash = Hash.new
text.each do
    |line|
    line =~ /^([A-Z]{2})\,([0-9\.]+)\,\-?([\.0-9]+) *$/
    state = $1
    latitude = $2
    longitude = $3
    lathash[state] = latitude.to_f
    lonhash[state] = longitude.to_f
end
text.close
img1 = Magick::ImageList.new("c\:\\ftp\\us\.gif")
width = img1.columns
height = img1.rows
gc = Magick::Draw.new
lathash.each do
    |key,value|
    state = key
    latitude = value.to_f
    longitude = lonhash[key].to_f
    offset_y = (((north - latitude) / (north - south)) * height).ceil
    offset_x = (((west - longitude) / (west - east)) * width).ceil
    gc.fill_opacity(0)
    gc.stroke('red').stroke_width(1)
    gc.circle(offset_x, offset_y, (offset_x - 15), (offset_y))
    gc.fill('black')
    gc.stroke('transparent')
    gc.text((offset_x - 5), (offset_y + 5), state)
    gc.draw(img1)
end
img1.border!(1,1, 'lightcyan2')
img1.write("us_out.gif")
exit
```

3.5.2 Analysis

The output is a U.S. map, without state borders, but with each state marked with a dot, at the location of its latitude and longitude (Figure 3.4).

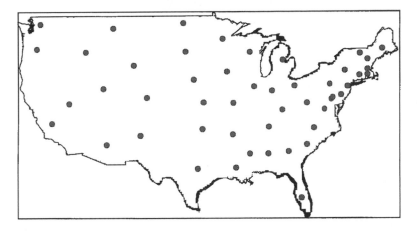

Figure 3.4 The output image, with circles at the coordinates of states.

Techniques whereby data files from one source are combined with maps, charts, or other visual tools are extremely useful. They permit us to examine complex data sources within a simple and familiar visual context. In later chapters, we will use data mashups to develop and test biomedical hypotheses.

Exercises

1. In Section 3.3 we created an image, but we did not display the image. Using Perl, Python, or Ruby, modify the script with Tk and display your created image in a widget.

2. Using your preferred language, and any image, "burn" your name into its lower-left corner of the image created in Exercise 1.

3. Using your preferred language, divide a photographic image (your choice) into 25 images of equal size.

4. Using any source you prefer (Google Earth, or any other map service), find the latitude and longitude of five hospitals. Modify the script to produce a map of the United States, marking the locations of these hospitals.

5. Using the data from Exercise 4, for at least one of the hospitals that you have found, find a county or state map that includes the hospital, and repeat Exercise 4, marking the location of the hospital within the state or county. You can see how this technique can track the disease occurrences of an epidemic if you know the latitude and longitudes of the homes containing individuals that contract the disease. It is now extremely easy to find the latitude and longitude of virtually every house in the United States and other countries.

4

INDEXING TEXT

A book index (Latin, from *indicare*, "to indicate") is a collection of significant words with page numbers as pointers to the location of those terms in the text. Likewise, our index finger is the pointer used to indicate the location of objects. A common misconception among some ebook enthusiasts is that the book index has been made obsolete by fast search features. In a typical ebook search, the user enters a term, and the computer scans the text until it encounters the first occurrence of the term. Each time the user presses the search button, the computer moves instantly to the next occurrence of the term. These searches are conducted without the benefit of an electronic index. When it is so easy and fast to do word or phrase searches, why would anyone want to build a book index?

Indexing, and a variety of related text organizing methods, are more important today than they have ever been, for reasons that have nothing to do with ebook searches. Text parsing searches that seem rapid for a 1 megabyte (MB) ebook will become excruciatingly slow when applied to a 1 terabyte data set produced by a hospital or a major healthcare agency. More importantly, searches on a single term will never suffice when the term has a dozen synonymous or related terms that must be included in a thorough review of the available data. Computational indexing techniques can automatically expand a term search into a term-relation search. Currently, the algorithms required to create detailed and comprehensive indices on very large data collections, are readily available. In this chapter, you will learn the rudiments of text indexing. We will use these fundamental algorithms again, in Chapter 14, when we discuss autocoding.

4.1 ZIPF Distribution of a Text File

In almost every segment of life, a small number of items usually account for the bulk of the observed activities. Though there are millions of authors, a relatively small number of authors account for the bulk of the books sold (think J.K. Rowling). A small number of diseases account for the bulk of deaths (think cardiovascular disease and cancer). A few phyla account for the bulk of the diversity of animals on earth (think arthropods). A few hundred words account for the bulk of all word occurrences in literature (think in, be, a, an, the, are). This phenomenon was observed and described by George Kingsley Zipf, who devised Zipf's law as a mathematical description.

Zipf's law applies to the diagnoses rendered in a pathology department. I helped write an early paper wherein 3 years' worth of a hospital's surgical pathology reports were collected and reviewed.

There were 64,921 diagnostic entries that were accounted for by 1,998 different morphologic diagnoses. A mere 21 diagnostic entities accounted for 50% of the different diagnoses collected by the pathology department. The data served to reassert Zipf's law, as it applies to pathology specimens.

You can create a Zipf distribution for any text file, listing the words that occur in the file, in descending order of their frequencies of occurrence.

4.1.1 Script Algorithm

1. Create a new file, meshzipf.txt, which will receive the output of the zipf distribution.
2. Obtain MeSH (Medical Subject Headings). In Chapter 5, we will be discussing the MeSH nomenclature, at some length. At this point, all we need to know about MeSH is that it is freely available as a large text file, from the following site:

   http://www.nlm.nih.gov/mesh/filelist.html

   Download the d2009.bin file (referred to as the ASCII MeSH download file). This plain-text file is about 28 MB in length and contains over 25,000 MeSH records. The record format of the d2009.bin file is described in the appendix.
3. Open the file, d2009.bin. On my computer, I've uploaded the file to my c:\big\ subdirectory, and the sample script is written to reflect this location.
4. Load the entire contents of the d2009.bin file into a string variable.
5. Parse the string variable, matching against each occurrence of a letter (uppercase or lowercase) followed by at least 2, and at most 15, lowercase letters, with the sequence bounded on either size by a word boundary. This pattern will bypass short words (one or two characters in length), or exceedingly long words. It will also exclude number, mixed alphanumerics, and words that are all-uppercase.
6. Create a dictionary object that will include words (keys) and number of occurrences (values).
7. Each time a word is matched by the regular expression, increment its number of occurrences by one.
8. After the dictionary object is complete (after the string variable containing the text has been parsed), format the values in the dictionary, as a zero-padded string of uniform length. This will permit the values to be sensibly sorted.
9. Sort the key–value pairs by values, ranging from the most frequently occurring word to the least frequently occurring word.
10. Print out the sorted value–key pairs to the meshzipf.txt output file.

Perl Script

```perl
#!/usr/local/bin/perl
open (OUT, ">meshzipf.txt");
open (TEXT, "c\:\\big\\d2009.bin");
undef($/);
$textvar = (<TEXT>);
while ($textvar =~ /(\b[A-Za-z][a-z]{2,15}\b)/g)
   {
   $freq{$1}++;
   }
$textvar = "";
while ((my $key, my $value) = each(%freq))
    {
    $value = "000000" . $value;
    $value = substr($value,-6,6);
    push (@termarray, "$value $key")
    }
@finalarray = reverse (sort (@termarray));
print OUT join("\n",@finalarray);
exit;
```

Python Script

```python
#!/usr/local/bin/python
import re
import string
word_list = []
freq_list = []
format_list = []
freq = {}
in_text = open('C:\\big\\d2009.bin', "r")
in_text_string = in_text.read()
out_text = open("meshzipf.txt", "w")
word_list = re.findall(r'(\b[A-Za-z][a-z]{2,15}\b)', in_text_string)
in_text_string = ""
for item in word_list:
  count = freq.get(item,0)
  freq[item] =  count + 1
for key, value in freq.iteritems():
  value = "000000" + str(value)
  value = value[-6:]
  format_list += [value + " " + key]
format_list = reversed(sorted(format_list))
print>>out_text, "\n".join(format_list)
exit
```

Ruby Script

```ruby
#!/usr/local/bin/ruby
freq = Hash.new(0)
```

```
zipfarray = []
file_out = File.open("meshzipf.txt","w")
file1 = File.open("c\:\\big\\d2009.bin", "r")
file1.read.scan(/\b[A-Za-z][a-z]{2,15}\b/){|word| freq[word] =
freq[word]+1}
file1 = ""
freq.each do
  |key, value|
  value = "000000" + value.to_s
  value = value.slice(-6,6)
  key = key.downcase
  zipfarray.push("#{value} #{key}")
end
zipfarray.sort.reverse.each {|item| file_out.puts item}
exit
```

4.1.2 Analysis

The first 10 entries from the zipf file for the MeSH file are:

   038979 the
   027634 and
   025435 abcdef
   025430 abbcdef
   017424 was
   015772 see
   015456 with
   012759 under
   010858 use
   010034 for

For these scripts, the entire content of a file is loaded into a string variable. The string variable is subsequently parsed into words, with each occurrence of the word counted. If the file is very large (exceeding the capacity of your computer's memory to process as a string), the script can be modified to read the file line by line, incrementing the word/frequency tally for the words contained in each line.

At the top of the Zipf list are the high-frequency words, such as "the", "and", and "was" that serve as connectors for lower-frequency, highly specific terms. Also included at the top of the Zipf list are frequently recurring letter sequences peculiar to the file; in this case, "abcdef" and "abbcdef". Zipf distributions have many uses in informatics projects, including the preparation of "stopword" lists (see Section 4.3).

4.2 Preparing a Concordance

A concordance is a special type of index, listing every location of every word in the text. Concordances can be used to support very fast proximity searches (finding the locations of words in proximity to other words), and phrase searches (finding sequences of words located in an ordered sequence somewhere in the text. Using only a concordance, it is a simple matter to computationally recreate the entire text. Preparing a concordance is quite simple.

4.2.1 Script Algorithm

1. Open a text file for reading. In this case, we open the file titles.txt, a collection of 100 public domain titles of journal articles. You can use any plain-text file for this script, or you can download titles.txt at

   http://www.julesberman.info/book/titles.txt

2. Read the entire contents of the file into a string variable.
3. Split the file into sentences. Rather than listing the page numbers where each word appears, we will be providing the sentence number for each appearance of each word in the text. We cannot provide page numbers for each word, because the text could be divided into pages in many different and arbitrary ways.
4. Parse each sentence into an array of words.
5. As each sentence is parsed, collect the words contained in the sentence and the location of the sentence. In this script, we indicate that the nth sentence of the file has a sentence location of n.
6. Add the location of the word to the dictionary object that contains the encountered words and their locations.
7. After the text has been parsed, order the words in the dictionary object alphabetically, and print out each word in the dictionary object, along with the dictionary value for each word (the locations where each word is found in the text).

Perl Script

```perl
#!/usr/local/bin/perl
open (TEXT, "titles.txt");
undef($/);
$textfile = <TEXT>;
$textfile =~ s/\n/ /g;
@sentencearray = split(/[\.\!\?] +(?=[A-Z])/, $textfile);
for ($i=0;$i<(scalar(@sentencearray));$i++)
    {
    $location = $i + 1;
    $sentence = lc($sentencearray[$i]);
```

```perl
    while ($sentence =~ /\b[a-z]{3,15}\b/g)
      {
      if (exists($wordhash{$&}))
        {
        $wordhash{$&} = $wordhash{$&} . "," . $location;
        }
      else
        {
        $wordhash{$&} = $location;
        }
      }
    }
foreach $key (sort(keys(%wordhash)))
  {
  print "$key $wordhash{$key}\n";
  }
exit;
```

Python Script

```python
#!/usr/local/bin/python
import re
import string
sentence_list = []
word_list = []
word_dict = {}
format_list = []
count = 0
stopfile = open("stop.txt",'r')
stop_list = stopfile.readlines()
stopfile.close()
in_text = open('titles.txt', "r")
in_text_string = in_text.read()
in_text_string = in_text_string.replace("\n"," ")
in_text_string = in_text_string.replace(" +"," ")
sentence_list = re.split(r'[\.\!\?] +(?=[A-Z])',in_text_string)
for sentence in sentence_list:
    count = count + 1
    sentence = string.lower(sentence)
    word_list = re.findall(r'(\b[a-z]{3,15}\b)', sentence)
    for word in word_list:
        if word_dict.has_key(word):
          word_dict[word] = word_dict[word] + ',' + str(count)
        else:
          word_dict[word] = str(count)
keylist = word_dict.keys()
keylist.sort()
for key in keylist:
```

```
  print key, word_dict[key]
exit
```

Ruby Script

```
#!/usr/local/bin/ruby
word_array = []
word_dict = {}
count = 0
text = IO.read("titles.txt")
text.gsub!(/\n+/," ") if text.include? "\n"
sentence_array = text.split(/[\.\!\?] +(?=[A-Z])/)
sentence_array.each do
 |sentence|
 count = count + 1
 sentence = sentence.downcase
 word_array = sentence.scan(/\b[a-z]{3,15}\b/)
 word_array.each do
   |word|
   count.to_s
   if word_dict.has_key?(word)
       word_dict[word] = "#{word_dict[word]}\,#{count}"
   else
       word_dict[word] = "#{count}"
   end
 end
end
out_array = word_dict.keys.sort
out_array.each{|word| puts "#{word} #{word_dict[word]}"}
exit
```

4.2.2 Analysis

The sample text consisted of 100 parsed sentences. Here are the first few lines of the output, consisting of words followed by the sentence numbers (ranging from sentence 1 to sentence 100), in which the word occurs.

ablation 32
ablative 47
absence 8
acetanilide 36
acid 24,37,44,71
acids 35
action 70
activation 6,19,20,35,46,57,95
activators 35

activity 14,36,58,63,71,75
acute 46
adaptive 27
adenocarcinoma 25
adenosine 52
adrenal 6
advanced 97
affects 69
after 29,32,38,46,79,97
against 43,61,71

4.3 Extracting Phrases

All text is composed of words and phrases that represent specific concepts, that are connected together into a sequence of meaningful statements.

Consider the following sentence:

"The diagnosis is chronic viral hepatitis."

This sentence contains two very specific medical concepts: "diagnosis" and "chronic viral hepatitis." The concepts are connected to form a meaningful statement with the words "the" and "is," and the sentence delimiter, ".".

"The," "diagnosis," "is," "chronic viral hepatitis," and "."

A phrase can be defined as a sequence of uncommon words that are terminated by the occurrence of a common word or by a sentence delimiter.

Here is another example:

"An epidural hemorrhage can occur after a lucid interval."

The medical concepts "epidural hemorrhage" and "lucid interval" are composed of uncommon words. These uncommon word sequences are bounded by sequences of common words or of sentence delimiters (the beginning or the end of a sentence).

"An," "epidural hemorrhage," "can occur after a," "lucid interval," and "."

If we had a list of all the words that were considered common, we could easily write a program that extracts all the concepts found in any text of any length. Common words are sometimes called "stopwords," because they mark the endings of concepts encountered in text.

The National Library of Medicine has published a list of public domain stopwords (Figure 4.1), available at:

http://www.nlm.nih.gov/bsd/disted/pubmedtutorial/020_170.html

This is a partial list. Click on Help in the text to see a full listing.

Stopwords

a	it	these
about	its	they
again	itself	this
all	just	those
almost	kg	through
also	km	thus
although	made	to
always	mainly	upon
among	make	use
an	may	used
and	mg	using
another	might	various
any	ml	very
are	mm	was
as	most	we
at	mostly	were

Figure 4.1   A partial listing of the National Library of Medicine's list of stopwords.

4.3.1 Script Algorithm

1. Open the stop.txt file, containing a list of (common) stopwords.
   Although most stopword lists will produce similar results, the stop.txt file, used in this script, is available at

   http://www.julesberman.info/book/stop.txt

2. Split the contents of the stop.txt file into a list structure.
3. Open the cancer_gene_titles.txt file. This file, prepared from a PubMed download and described in Chapter 9, is available as a public domain text file at:

   http://www.julesberman.info/book/cancer_gene_titles.txt

   The file contains over 18,000 titles of scientific articles, without punctuation, one title per line.

4. Parse through the lines of the text. Substitute a newline character for every occurrence of any stopword in the sentence.
5. Split the resulting sentences at every newline character contained in the sentence. The split fragments of each sentence comprise the concepts contained in the sentence.

6. Put the concepts into a data structure that accumulates the concepts for each sentence in the text.

7. Remove duplicate items, sort the remaining items alphabetically, and print them to an output file.

Perl Script

```perl
#!/usr/local/bin/perl
open (STOPFILE, "stop.txt");
undef($/);
@stoparray = split(/\n/, <STOPFILE>);
$/ = "\n";
open (TEXT, "cancer_gene_titles.txt");
open (OUT, ">phrases.txt");
$sentence = " ";
while ($sentence ne "")
    {
    $sentence = <TEXT>;
    $sentence =~ s/\n$/ /;
    foreach $stopword (@stoparray)
      {
      $sentence =~ s/ *\b$stopword\b */\n/g;
      }
    push(@phrasearray, (split(/\n/, $sentence)));
    }
%stophash = map{$_, ""} @phrasearray;
@phrasearray = sort(keys(%stophash));
print OUT join("\n",@phrasearray);
exit;
```

Python Script

```python
#!/usr/local/bin/python
import re, string
item_list = []
stopfile = open("stop.txt",'r')
stop_list = stopfile.readlines()
stopfile.close()
in_text = open('cancer_gene_titles.txt', "r")
count = 0
for line in in_text:
  count = count + 1
  for stopword in stop_list:
    stopword = re.sub(r'\n', '', stopword)
    line = re.sub(r' *\b' + stopword + r'\b *', '\n', line)
  item_list.extend(line.split("\n"))
item_list = sorted(set(item_list))
out_text = open('phrases.txt', "w")
```

```
for item in item_list:
  print>>out_text, item
exit
```

Ruby Script

```ruby
#!/usr/local/bin/ruby
phrase_array = []
stoparray = IO.read("stop.txt").split(/\n/)
sentence_array = IO.read("cancer_gene_titles.txt").split(/\n/)
out_text = File.open("phrases.txt", "w")
sentence_array.each do
  |sentence|
  stoparray.each do
    |stopword|
    sentence.gsub!(/ *\b#{stopword}\b */, "\n") if sentence.include?
stopword
  end
  phrase_array = phrase_array + sentence.split(/\n/)
end
out_text.puts phrase_array.sort.uniq
exit
```

4.3.2 Analysis

The output is an alphabetic file of the phrases that might appear in a book's index. A sampling of the output is shown in Figure 4.2. We used the file consisting of titles from a PubMed search. This file, cancer_gene_titles.txt, is about 1.1 MB in length, the size of a typical book. We only required about a dozen lines of code and a few seconds of execution time to create our list of index terms. Creating the final index will require us to visually read the list, excluding unhelpful terms. Afterwards, we can write a short script that assigns page numbers to the final list of index terms.

4.4 Preparing an Index

An index is a list of the important words or phrases contained in a book, along with the locations where each of those words and phrases can be found.

An index differs from a concordance because the index does not contain every word found in the text, and the index contains groups of selected phrases, in addition to individual words.

A good index cannot be created exclusively by a software program. Although it would be easy to write a script that finds every word and every sequence of words in a text, and produces a file that lists all of these strings, and their locations, the resulting product would have a length many times that of the original book.

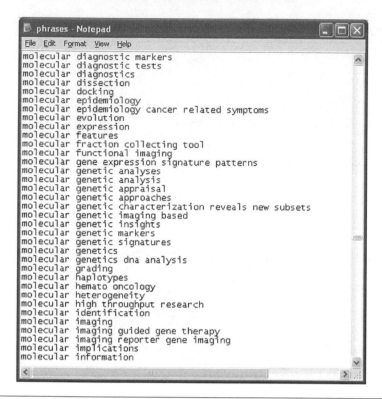

Figure 4.2 A small sampling of the text phrases extracted with the stopword method.

For example, let's pretend that we would like to publish a book that is only one sentence in length and consists of the following text: "This book is short." Here is the complete index, listing each word and phrase and the word number in the sentence where the word or phrase is found, in alphabetic order.

book – 2
book is – 2
book is short – 2
is – 3
is short – 3
short – 4
this – 1
this book – 1
this book is – 1
this book is short – 1

A complete index is always much longer than the length of a book. A useful index is selective, containing only those words and phrases that would be of greatest interest to the reader.

4.4.1 Script Algorithm

1. Create an array containing stopwords. You can use any stopword list you prefer. In this script, we use stop.txt available at http://www.julesberman.info/book/stop.txt

2. Open a file to be indexed. You can use any file, but in this text, we use text.txt, available at http://www.julesberman.info/book/text.txt

3. Strip the text of any non-ASCII characters (not necessary if you are using a plain-text file).

4. Split the text into sentences and put the consecutive sentences into an array.

5. Create a dictionary object, which will hold phrases as keys and a comma-separated list of numbers, representing the sentences in which the phrases appear, as the values.

6. For each sentence in the array of consecutive sentences, split the sentence wherever a stopword appears, and put the resulting phrases into an array.

7. For each array of phrases, from each sentence, parse through the array of phrases, assigning each phrase to a dictionary key, and concatenating the sentence number in which the phrase occurs, to the comma-separated list of sentence numbers that serves as the value for the key (phrase)

Perl Script

```
#!/usr/local/bin/perl
open (STOPFILE, "stop.txt");
open (OUT, ">index.out");
undef($/);
@stoparray = split(/\n/, <STOPFILE>);
open (TEXT, "text.txt");
$textfile = <TEXT>;
$textfile =~ s/\n/ /g;
@sentencearray = split(/[\.\!\?] +(?=[A-Z])/, $textfile);
for ($i=0;$i<(scalar(@sentencearray));$i++)
    {
    $location = $i + 1;
    $sentence = lc($sentencearray[$i]);
    $sentence =~ s/\'s//;
    $sentence =~ tr/\000-\011\013-\014\016-\037\041-\055\173-\377//d;
    $sentence =~ s/^ +//;
    $sentence =~ s/ +$//;
    $sentence =~ s/ +/ /;
    foreach $stopword (@stoparray)
       {
       $sentence =~ s/\b$stopword\b/\*/g;
       }
    @phrasearray = split(/ *\* */, $sentence);
    foreach $phrase (@phrasearray)
       {
       if (exists($phrasehash{$phrase}))
```

```perl
         {
         $phrasehash{$phrase} = $phrasehash{$phrase} . "," . $location;
         }
       else
         {
         next if ($phrase eq "");
         next if ($phrase =~ /^[^a-z]/);
         $phrasehash{$phrase} = $location;
         }
       }
     }
foreach $key (sort(keys(%phrasehash)))
  {
  print OUT "$key $phrasehash{$key}\n";
  }
exit;
```

Python Script

```python
#!/usr/local/bin/python
import re
import string
item_list = []
item_dictionary = {}
place_string = ""
stopfile = open("stop.txt",'r')
stop_list = stopfile.readlines()
stopfile.close()
in_text = open('text.txt', "r")
in_text_string = in_text.read()
in_text_string = in_text_string.replace("\n"," ")
in_text_string = in_text_string.replace(" +"," ")
sentence_list = re.split(r'[\.\!\?] +(?=[A-Z])',in_text_string)
norm = string.maketrans('','')
badascii = string.translate(norm,norm,string.printable)
badascii_table = badascii + (256 - len(badascii))*" "
junk_table = 256*" "
table = string.maketrans(badascii_table,junk_table)
count = 0
for item in sentence_list:
    count = count + 1
    count_string = str(count)
    item = string.lower(item)
    item = re.sub(r'\'s', "", item)
    item = item.translate(table)
    for stopword in stop_list:
        stopword = string.rstrip(stopword)
        item = re.sub(r' *\b' + stopword + r'\b *', '\n', item)
```

```
      item_list = item.split("\n")
      for phrase in item_list:
          phrasematch = re.match(r'^[0-9]', phrase)
          if (phrasematch):
              continue
          if item_dictionary.has_key(phrase):
              item_dictionary[phrase] = item_dictionary[phrase] + ',' + ⮐
count_string
          else:
              item_dictionary[phrase] = count_string
keylist = item_dictionary.keys()
keylist.sort()
for key in keylist:
  print key, item_dictionary[key]
exit
```

Ruby Script

```ruby
#!/usr/local/bin/ruby
phrase_array = []
phrasehash = {}
stoparray = IO.read("stop.txt").split(/\n/)
text = IO.read("text.txt")
text.gsub!(/\n+/," ") if text.include? "\n"
sentence_array = text.split(/[\.\!\?] +(?=[A-Z])/)
sentence_array.each do
 |sentence|
 count = sentence_array.rindex(sentence) + 1
 sentence = sentence.downcase.tr('^a-z0-9 ',' ').strip.gsub(/ +/," ⮐
")
 stoparray.each do
   |stopword|
   stopword = stopword.chomp
   sentence.gsub!(/ *\b#{stopword}\b */, "\n") if sentence.include?
stopword
   end
 sentence.split(/\n/).each do
   |phrase|
   next if phrase == ""
   next if phrase =~ /^[^a-z]/
   if phrasehash.has_key?(phrase)
     phrasehash[phrase] = "#{phrasehash[phrase]}\, #{count}"
   else
     phrasehash[phrase] = count
   end
 end
end
phrasehash.keys.sort.each {|key| puts "#{key} #{phrasehash[key]}"}
exit
```

4.4.2 Analysis

An example of the kind of output produced by the script is shown:

    adjustment 7,9
    adjuvant chemotherapy 83
    adjuvant imrt 23
    adjuvant treatment 10
    administered 82
    adult 73
    advanced stages 71
    age 9
    agree 52
    aim 46,73,81
    analysed 58
    analyzed 48
    anaplastic histology 24
    apaap technique 85
    april 1982 54
    aspiration 86
    associated obstructive jaundice 76
    asymptomatic twin 43,44
    attention 21
    avoided 79

The numbers represent the sentence numbers in which each phrase occurs. Had we been preparing an index for a book, we would have written the script to capture page numbers. In a real index, we would probably be selective, deleting obvious or irrelevant entries. We might also wish to group related entries by indentation.

For example:

    adjuvant
    adjuvant chemotherapy 83
    adjuvant imrt 23
    adjuvant treatment 10

We might have combined equivalent words that have minor spelling variations. For example:

    analyzed (var. analysed) 48, 58

Automatic indexing invariably produces a product that a human indexer can improve. The strength of automatic indexing is found when the texts are very long (gigabytes or greater). Humans simply cannot index long texts. A flawed computer-generated index is usually better than no index at all.

4.5 Comparing Texts Using Similarity Scores

When you have extracted all of the phrases occurring in a text, as we did in the prior section, you have created something akin to the signature of the text. The phrases that comprise the text tell us a great deal about the contents of the text—more so than the collection of included words. A phrase is a description in miniature, and contains a sequence of words, in a very definite order, that characterizes the way that the author expresses his or her ideas. It would be very unlikely for two different documents to have the same phrase list, and unlikelier still to have the same phrase list, with each phrase occurring at the same frequency in the two different documents.

We can use the list of phrases occurring in a text, along with the frequencies of occurrence of those phrases, to create a type of signature that identifies and describes the full text. We can determine whether two different texts are similar, when we compare their signatures.

Similarity scores are very useful in the medical sciences. We use similarity scores to establish relatedness of objects (e.g., DNA sequences), to find trends and outliers in population data, to provide "best-fit" search results, and to classify groups of items (e.g., cluster analysis). These methods usually begin with calculations of the pairwise similarity scores among objects in a population.

Often, the similarity score is based on comparing some set of measured features for two different objects, and summing the squares of the differences in magnitude for each measured feature. One such scoring system is the Pearson correlation, which produces a score that can vary from –1 to +1. Two objects with a high score (near +1) are highly similar. In this section, we will determine the Pearson score of two books. The measured feature of the books will be the list of phrases contained in the books, and the number of their occurrences.

4.5.1 Script Algorithm

1. We could compare any two documents, but for this exercise we chose Stevenson's *Treasure Island* and Milton's *Paradise Lost*. The two novels represent very different writing styles (gruff prose versus ornate verse), and contain disparate vocabularies. The etext versions of these books are in the public domain and can be downloaded from Project Gutenberg at the following URLs:

   For *Paradise Lost*

   http://www.gutenberg.org/dirs/etext91/plboss10.txt

   For *Treasure Island*

   http://www.gutenberg.org/etext/120

Name the two downloaded files "paradise.txt" and "treasure.txt", respectively, and put them into the same subdirectory containing your script.

2. Put the names of each text file into an array. We will be performing the same parsing steps on each of the two files.

3. Open the stop.txt file, containing the high-frequency stopwords that we will use to determine the boundaries of a phrase. (*Remember*: An index phrase is a sequence of words bounded on both sides by a stop word or by the beginning or the end of a sentence.) The stop file consists of one word per file line. Put all of the words from the stop.txt file into an array, stripping the newline character that separates each stop word from the subsequent stop word.

4. Open the first text file (*Paradise Lost*), and read the entire text into a string variable.

5. Delete every newline character from the text file string, replacing it with a space character.

6. In the text file string, wherever there is a sequence of words bounded on either side by a stopword, replace the stopwords with a newline character. Iterate this determination and replacement, over the entire text file string, for every stopword in our array of stop words.

7. Wherever there is a ",", ":", ";", "(" or ")" in the text file string, replace the punctuation with a newline character. We do this because these punctuation marks delineate the beginning and the end of an expression and, for the purposes of delineating index phrases, these punctuation marks are equivalent to an end-of-sentence marker.

8. Wherever the text file string has a ".", or "!" or "?" followed by one or more spaces, followed by an uppercase letter, replace the punctuation and the following white spaces with a newline character. We do this because the pattern is typical of a sentence ending, and sentence endings mark the end of index phrases.

9. Convert the modified text file string, which now marks the beginning and ending of index phrases with newline characters, into lowercase.

10. Convert the modified text file string, replacing all sequences consisting of multiple space characters with a single space character.

11. Split the text file string into an array, at every occurrence of a newline character bordered by zero or more spaces. This results in an array that includes all of the index phrases in the original text file.

12. Iterate through every phrase in the newly created array of index phrases.

13. For each phrase, if the phrase does not match a sequence of lowercase letters followed by a space followed by a sequence of lowercase letters or spaces, skip to the next item in the phrase array. We do this primarily to eliminate single word phrases that do not contain a space intervening between words. This step also eliminates phrases that contain numeric and nonalphabet characters.

14. We will be using two dictionary objects: the dictionary object consisting of all of the index phrases from *Paradise Lost* as keys, and the number of occurrences

of each index phrase in *Paradise Lost* as the values, as well as the index phrases that occur exclusively in *Treasure Island*, all with the number "0" as the value. The other dictionary object will consist of the index phrases from *Treasure Island* as keys, and the number of occurrences of each index phrase from *Treasure Island*, as the values, as well as the index phrases that occur exclusively in *Paradise Lost*, all with the number "0" as the value. By creating these two dictionary objects, we create two dictionary objects that have the same matching set of keys, with one set of keys holding the number of occurrences of the keys in *Paradise Lost*, and the other holding the number of occurrences of the keys in *Treasure Island*. We can then compare each dictionary object key by key and value by value.

15. To create the two dictionary objects, increment each occurrence of a phrase by one in the dictionary object for the text file in which it has occurred, and create a key–value pair in the other text file's dictionary object (if none exists) consisting of the phrase and the value "0".

16. Repeat steps 4 to 15 for the second book, *Treasure Island*. When you have repeated these steps for the second book you will have collected the two dictionary objects that you will use to compute the Pearson score. At this point, you could substitute any similarity correlation scores you prefer over the Pearson score.

17. Parse over every key–value pair in either dictionary object (we chose the dictionary object for *Paradise Lost*, but the calculation, which depends on differences between the two dictionary objects, would yield the same score using either dictionary object).

18. Keep a count of the total number of key–value pairs.

19. Produce a summation tally of the values in the *Paradise Lost* dictionary object and in the *Treasure Island* dictionary object.

20. Produce a summation tally of the squares of the values in the *Paradise Lost* dictionary object and the squares of the values in the *Treasure Island* dictionary object.

21. Produce a summation tally of the products of each value in the *Paradise Lost* dictionary object multiplied by the corresponding value (the value of the same key) in the *Treasure Island* dictionary object.

22. After the dictionary object is parsed, you will take the tally variables that you just computed, and you will insert them into the Pearson formula.

23. The Pearson score is the summation tally of the products minus the sum tally of the first dictionary object times the sum tally of the second dictionary object divided by the number of keys in the object all divided by the square root of the tally of the squares of the values of the *Paradise Lost* dictionary object times the square of the sum tally of *Paradise Lost* dictionary object divided by the number of keys in the object, times the tally of the squares of the values of the *Treasure Island* dictionary object times the square of the sum tally of *Treasure Island* dictionary object divided by the number of keys in the object.

Step 23 is an example where the description of a mathematical expression, in English, is much, much more confusing than the program code for the mathematical expression.

Perl Script

```perl
#!/usr/local/bin/perl
@filearray = ("paradise.txt", "treasure.txt");
undef($/);
foreach $filename (@filearray)
  {
  open (STOPFILE, "c\:\\ftp\\stop.txt");
  @stoparray = split(/\n/, <STOPFILE>);
  open (TEXT, "$filename")||die"cannot";
  $textfile = <TEXT>;
  $textfile =~ s/\n/ /g;
  foreach $stopword (@stoparray)
      {
      $textfile =~ s/\b$stopword\b/\n/g;
      $textfile =~ s/[\,\:\;\(\)]/\n/g;
      }
  $textfile =~ s/[\.\!\?] +(?=[A-Z])/\n/g;
  $textfile = lc($textfile);
  $textfile =~ s/ +/ /g;
  @phrasearray = split(/ *\n */, $textfile);
  foreach $phrase (@phrasearray)
    {
    $phrase =~ s/ +/ /;
    $phrase =~ s/^ +//;
    $phrase =~ s/ +$//;
    next if ($phrase !~ /^[a-z]+ [a-z ]+$/);
    if ($filename eq "paradise.txt")
      {
      $paradise{$phrase}++;
      $treasure{$phrase} = 0 unless exists($treasure{$phrase});
      }
    if ($filename eq "treasure.txt")
      {
      $treasure{$phrase}++;
      $paradise{$phrase} = 0 unless exists($paradise{$phrase});
      }
    }
  close TEXT;
  }
while ((my $key, my $value) = each(%paradise))
  {
  $count++;
  $sumtally1 = $sumtally1 + $value;
  $sumtally2 = $sumtally2 + $treasure{$key};
```

```perl
    $sqtally1 = $sqtally1 + $value**2;
    $sqtally2 = $sqtally2 + $treasure{$key}**2;
    $prodtally12 = $prodtally12 + ($value * $treasure{$key});
    }
$part1 = $prodtally12 - ($sumtally1 * $sumtally2 / $count);
$part2 = $sqtally1 - (($sumtally1)**2 / $count);
$part3 = $sqtally2 - (($sumtally2)**2 / $count);
$similarity12 = $part1 / sqrt($part2 * $part3);
print "The Pearson score is " . $similarity12 . "\n";
exit;
```

Output:

C:\>perl pearson.pl

The Pearson score is −0.477744298690063

Python Script

```python
#!/usr/local/bin/python
import re
import string
from math import sqrt
from math import pow
treasure = {}
paradise = {}
filelist = ["treasure.txt", "paradise.txt"]
stopfile = open("stop.txt",'r')
stop_list = stopfile.readlines()
stopfile.close()
phraseform = re.compile(r'^[a-z]+ [a-z ]+$')
for filename in filelist:
    in_text = open(filename, "r")
    in_text_string = in_text.read()
    in_text.close()
    in_text_string = in_text_string.replace("\n"," ")
    for stopword in stop_list:
      stopword = string.rstrip(stopword)
      in_text_string = re.sub(r' *\b' + stopword + r'\b *',
'\n',in_text_string)
    in_text_string = re.sub(r'[\,\:\;\(\)]','\n',in_text_string)
    in_text_string = re.sub(r'[\.\!\?] +(?=[A-Z])', '\n',
in_text_string)
    in_text_string = string.lower(in_text_string)
    item_list = re.split(r' *\n *', in_text_string)
    for phrase in item_list:
        phrase = re.sub(r' +',' ', phrase)
        phrase = string.strip(phrase)
        phrasematch = phraseform.match(phrase)
        if not (phrasematch):
            continue
```

```
       if (filename == "paradise.txt"):
         if paradise.has_key(phrase):
           paradise[phrase] = paradise[phrase] + 1
         else:
           paradise[phrase] = 1
         if not (treasure.has_key(phrase)):
           treasure[phrase] = 0
       if (filename == "treasure.txt"):
         if treasure.has_key(phrase):
           treasure[phrase] = treasure[phrase] + 1
         else:
           treasure[phrase] = 1
         if not (paradise.has_key(phrase)):
           paradise[phrase] = 0
count = 0; sumtally1 = 0; sumtally2 = 0; sqtally1 = 0; sqtally2 = 0
prodtally12 = 0; part1 = 0; part2 = 0; part3 = 0;
keylist = paradise.keys()
for key in keylist:
  count = count + 1;
  sumtally1 = sumtally1 + paradise[key]
  sumtally2 = sumtally2 + treasure[key]
  sqtally1 = sqtally1 + pow(paradise[key],2)
  sqtally2 = sqtally2 + pow(treasure[key],2)
  prodtally12 = prodtally12 + (paradise[key] * treasure[key])
part1 = prodtally12 - (float(sumtally1 * sumtally2) / count)
part2 = sqtally1 -  (float(pow(sumtally1,2)) / count)
part3 = sqtally2 -  (float(pow(sumtally2,2)) / count)
similarity12 = float(part1) / float(sqrt(part2 * part3))
print "The Pearson score is", similarity12
exit
```

Output:

    C:\>python pearson.py

The Pearson score is −0.47774429869

The Python script requires importation of three external modules: re, string, and math. The math module supports the sqrt (square root function) and the pow (exponentiation) function.

Ruby Script

```
#!/usr/local/bin/ruby
require 'mathn'
treasure = {}; paradise = {};
filelist = ["treasure.txt", "paradise.txt"]
filelist.each do
  |filename|
  stoparray = IO.read("stop.txt").split(/\n/)
```

```ruby
    text = IO.read(filename)
    text.gsub!("\n"," ") if text.include? "\n"
    stoparray.each do
      |stopword|
      stopword = stopword.chomp
      text.gsub!(/ *\b#{stopword}\b */, "\n") if text.include?
stopword
      text.gsub!(/[\,\:\;\(\)]/, "\n") if text =~ /[\,\:\;\(\)]/
    end
    text.gsub!(/[\.\!\?] +(?=[A-Z])/, "\n") if text =~ /[\.\!\?]
+(?=[A-Z])/
    text.downcase.strip.split(/ *\n */).each do
      |phrase|
      phrase.gsub!(/ +/," ") if phrase =~ / +/
      phrase.strip!
      next if phrase !~ /^[a-z]+ [a-z ]+$/
      if filename == "paradise.txt"
        if paradise.has_key?(phrase)
          paradise[phrase] = paradise[phrase] + 1
        else
          paradise[phrase] = 1
        end
        treasure[phrase] = 0 unless treasure.has_key?(phrase)
      end
      if filename == "treasure.txt"
        if treasure.has_key?(phrase)
          treasure[phrase] = treasure[phrase] + 1
        else
          treasure[phrase] = 1
        end
        paradise[phrase] = 0 unless paradise.has_key?(phrase)
      end
    end
end
count = 0; sumtally1 = 0; sumtally2 = 0; sqtally1 = 0; sqtally2 = 0
prodtally12 = 0; part1 = 0; part2 = 0; part3 = 0;
paradise.each do
  |k,v|
  count = count + 1;
  sumtally1 = sumtally1 + v
  sumtally2 = sumtally2 + treasure[k]
  sqtally1 = sqtally1 + v**2;
  sqtally2 = sqtally2 + treasure[k]**2
  prodtally12 = prodtally12 + (v * treasure[k])
end
part1 = prodtally12 - ((sumtally1 * sumtally2).to_f / count.to_f)
part2 = sqtally1 -  (((sumtally1)**2).to_f / count.to_f)
part3 = sqtally2 -  (((sumtally2)**2).to_f / count.to_f)
```

```
similarity12 = part1.to_f / (Math.sqrt(part2 * part3)).to_f
puts "The Pearson score is #{similarity12}"
exit
```

Output:

C:\>ruby pearson.rb

The Pearson score is −0.477744298690063

The Ruby script requires importation of the mathn module, to support high precision division, and uses the Math module's sqrt (square root) method.

4.5.2 Analysis

Pearson scores range from −1 to 1. A score of 1 occurs when a document is compared against itself. When we compute the Pearson score between two highly dissimilar texts, the yielded score is −0.4777. We expected and received a low-end Pearson score.

Some correlation tests impose severe requirements on the set of values compared in different objects. For example, a correlation test might require that each of two objects have the same number and type of compared items, and that items must be associated with a value that is confined to the same range in both objects. For example, if comparing test scores for two students, you might need to have each student take the same set of tests, with each test graded by the same grader, applying a test score within the same range (e.g., 50 to 100). The Pearson score is popular because it tolerates a wide range of disparities between the two correlated objects. In this example, we used two texts that had different lengths, different words, and a wide range in the occurrence value for the different words. The Pearson score compensates, producing a score that seems to have some validity.

If you doubt the utility of the Pearson score, you can test it yourself, using highly similar documents or dissimilar documents. To create two highly similar documents, you can take one document and truncate it by half, determining the Pearson score of the entire document against the first half of the document, or scoring the first half of the document against the second half of the document. If you're still unsatisfied, feel free to develop your own similarity score. For a specific type of document, it is quite possible to devise a scoring system that is superior to the Pearson score.

Exercises

1. The script that extracts the different words of a text, and determines the frequency of each word, begins by gobbling the entire file into a variable. If the file size exceeds the memory constraints for string variables, the script will not execute. Modify the script so that it parses the file in memory-tolerant parts,

building the list of different words, and their frequencies, with each file-read operation, until the entire file, regardless of its length, is parsed.

2. Using Perl, Python, or Ruby, prepare a phrase list for OMIM.

3. Our script that produced a Zipf distribution of the MeSH file, produced an output wherein each word in the MeSH file was listed, along with the number of occurrences in the file, in decreasing number of occurrences. Using Perl, Python or Ruby, modify or rewrite the script to produce a cumulative tally of the percentage of all occurrences accounted for by words preceding and including each line in the output.

   For example, examine the Zipf distribution of the following 10-word sentence: "When I am here, I am not there, but here."

   The Zipf distribution is

   I 2
   am 2
   here 2
   but 1
   not 1
   there 1
   When 1

   The cumulative tally is

   I 20%
   am 40%
   here 60%
   but 70%
   not 80%
   there 90%
   When 100%

   How many words account for 50% of the word occurrences in the MeSH file?

4. The script that produces a concordance of a text file has several small bugs. It will exclude words that contain an apostrophe (e.g., Hodgkin's). It will redundantly include sentence locations for words that occur more than once within the sentence. Using your preferred language, rewrite the script to correct these bugs.

5. There are many ways of computationally narrowing the number of important words and phrases within a book. The easiest method is to select only those phrases that fall into an allowed length (e.g., three words or less) or that occur rarely (e.g., fewer than 10 times within the book), or that are composed of words that do not occur in other phrases. Using any refinement method you prefer (not necessarily from the suggestions above) prepare a script (in your favorite language) that produces an index of reduced size by filtering out undesirable terms.

6. Using Perl, Python, or Ruby, create your own similarity method to replace the Pearson method. You can do this by modifying the Pearson method, or by devising a totally different approach to similarity scoring.

   Hint: A similarity method may combine several different comparison metrics for a document (number of bytes of document, average length of a sentence in the document, degree of overlap of vocabulary in the document, most frequently occurring word) and producing a score that somehow combines all of these values.

7. When you have a satisfactory way of measuring the similarity between two things, the next step is to group the members of a set of items, based on similarity. This is sometimes called "clustering." A great many different clustering algorithms have been produced, but each algorithm begins with choosing a way of measuring the similarity between items in the group. Prepare a hierarchical algorithmic scheme that performs repeated similarity scores on the members of a group of items (e.g., documents), until each item is assigned into a cluster of similar items. You can develop your own algorithm, or you can reproduce an algorithm selected from the scientific literature.

PART II
MEDICAL DATA
RESOURCES

5

THE NATIONAL LIBRARY OF MEDICINE'S MEDICAL SUBJECT HEADINGS (MeSH)

Nomenclatures are comprehensive repositories of domain terminologies. Moreover, modern nomenclatures are keys to all the knowledge pertaining to any of the terms in the nomenclature. A well-organized, comprehensive nomenclature can be used to annotate and index any information in any document, and permit that information to be retrieved and merged with relevant information contained in other documents. Under ideal conditions, a nomenclature creates new knowledge by exploiting the relationships among terms that annotate the biomedical literature.

MeSH (Medical Subject Headings) is a wonderful nomenclature of medical terms available from the U.S. National Library of Medicine.

The download site is

http://www.nlm.nih.gov/mesh/filelist.html

The recent ASCII MeSH download file is

d2009.bin (27,369,460 bytes)

MeSH contains over 25,000 records. The first record in MeSH is shown in Figure 5.1.

MeSH is one of the greatest gifts provided by the U.S. National Library of Medicine and can be used freely for a variety of projects involving indexing, tagging, searching, retrieving, coding, analyzing, merging, and sharing biomedical text. In my opinion, there are many projects that rely on commercial and legally encumbered nomenclatures that would be better served by MeSH.

My only quibble with MeSH is that it is incorrectly described as a tree structure.

Here is the official word (from the NLM Web site) on MeSH trees from http://www.nlm.nih.gov/mesh/intro_trees2007.html: "Because of the branching structure of the hierarchies, these lists are sometimes referred to as 'trees.' Each MeSH descriptor appears in at least one place in the trees, and may appear in as many additional places as may be appropriate. Those who index articles or catalog books are instructed to find and use the most specific MeSH descriptor that is available to represent each indexable concept."

When you look at individual entries in MeSH, you find that a single entry may be assigned multiple MeSH numbers.

```
*NEWRECORD
RECTYPE = D
MH = Calcimycin
AQ = AA AD AE AG AI AN BI BL CF CH CL CS CT DU EC HI IM IP ME PD PK PO RE SD ST TO TU UR
ENTRY = A-23187|T109|T195|LAB|NRW|NLM (1991)|900308|abbcdef
ENTRY = A23187|T109|T195|LAB|NRW|UNK (19XX)|741111|abbcdef
ENTRY = Antibiotic A23187|T109|T195|NON|NRW|NLM (1991)|900308|abbcdef
ENTRY = A 23187
ENTRY = A23187, Antibiotic
MN = D03.438.221.173
PA = Anti-Bacterial Agents
PA = Ionophores
MH_TH = NLM (1975)
ST = T109
ST = T195
N1 = 4-Benzoxazolecarboxylic acid, 5-(methylamino)-2-((3,9,11-trimethyl-8-(1-methyl-2-oxo-2-(1H-pyrrol-2-yl)ethyl)-1,7-
dioxaspiro(5.5)undec-2-yl)methyl)-, (6S-(6alpha(2S*,3S*),8beta(R*),9beta,11alpha))-
RN = 52665-69-7
PI = Antibiotics (1973-1974)
PI = Carboxylic Acids (1973-1974)
MS = An ionophorous, polyether antibiotic from Streptomyces chartreusensis. It binds and transports cations across membranes and
uncouples oxidative phosphorylation while inhibiting ATPase of rat liver mitochondria. The substance is used mostly as a biochemical
tool to study the role of divalent cations in various biological systems.
OL = use CALCIMYCIN to search A 23187 1975-90
PM = 91; was A 23187 1975-90 (see under ANTIBIOTICS 1975-83)
HN = 91(75); was A 23187 1975-90 (see under ANTIBIOTICS 1975-83)
MED = *62
MED = 847
M90 = *299
M90 = 2405
M85 = *454
M85 = 2878
M80 = *316
M80 = 1601
M75 = *300
M75 = 823
M66 = *1
M66 = 3
M94 = *153
M94 = 1606
MR = 20060705
DA = 19741119
DC = 1
DX = 19840101
UI = D000001
```

Figure 5.1   The first record in the MeSH data file.

For example, the MeSH term "Family" is assigned two MeSH numbers:

MN = F01.829.263
MN = I01.880.225

The parent "number" for any MeSH entry is found by removing the last set of decimal demarcated digits.

For example:

F01.829.263 MeSH name, Family
F01.829 MeSH name, Psychology, Social
F01 MeSH name, Behavior and Behavior Mechanisms

For each MeSH number, there is a separate hierarchy.

It is tempting to think of each hierarchy for each number as a tree (then MeSH could be envisioned as a dense forest), but each parent term could be assigned multiple MeSH numbers, each producing a multibranching hierarchy.

Because each MeSH term (including the ancestral terms for a MeSH term) may be assigned multiple MeSH numbers, each with its own hierarchy, the MeSH data structure is more accurately thought of as a complex ontology, with terms existing in multiple classes, with specified relationships among any class and its parent classes.

The tree metaphor breaks down because branches and nodes within a branch can be connected to other branches and to other nodes. Trees do not do this kind of thing.

5.1 Determining the Hierarchical Lineage for MeSH Terms

It is possible to write a script that parses through every MeSH entry, finds all of the MeSH numbers for the entry, determines the parent terms for the MeSH numbers, determines all of the alternate MeSH numbers for the parent terms, then finds all of the grandparent terms for all of the parent terms, etc., until all of the hierarchical terms for the term are found.

5.1.1 Script Algorithm

1. Open the ASCII version of the MeSH file, for reading.
2. Open a file for writing. This file will receive the full hierarchy of the MeSH terms.
3. Parse through the MeSH file, line by line.
4. When a line that begins with "MH = " occurs, capture the MeSH term found on the line.
5. When a line that begins with "MN = " occurs, capture the MeSH number found on the line.
6. Because the MeSH term always occurs before the MeSH number, the MeSH number will always be the number that corresponds with the previously captured MeSH term. Use the captured number and the captured term to create a new key–value pair for a dictionary object.
7. Because a single MeSH term may have several different MeSH numbers listed in its record, as additional lines are encountered, concatenate the list of MeSH numbers for a single term into a string.
8. Create another dictionary object with key–value pairs consisting of a MeSH term key and the concatenated collection of all corresponding MeSH numbers, as the value.
9. When the entire MeSH ASCII file has been parsed, close the file.
10. Parse through the collection of key–value pairs in the dictionary object containing the MeSH terms as keys and the MeSH numbers as values. For each key–value pair, repeat steps 12–15.
11. Print each key as a line on an external file.
12. Create an array consisting of the different MeSH numbers corresponding to the value of the key.

13. For each mesh number, chop the item into an array of MeSH numbers consisting of iterative truncations of the original mesh number, at the decimal points in the MeSH number. This produces a list of the parent numbers of the MeSH number.

14. Alphabetize the MeSH numbers in the array, and print out each MeSH number, followed by its corresponding MeSH term.

Perl Script

```perl
#!/usr/local/bin/perl
open(MESH, "C\:\\BIG\\D2009.BIN");
open(OUT, ">mesh.out");
$line = " ";
while ($line ne "")
  {
  $line = <MESH>;
  $name = $1 if ($line =~ /MH = (.+)$/);
  if ($line =~ /MN = (.+)$/)
      {
      $number = $1;
      $numberhash{$number} = $name;
      if (exists($namehash{$name}))
        {
        $namehash{$name} = $namehash{$name} . " " . $number;
        }
      else
        {
        $namehash{$name} = $number;
        }
      }
  }
close(MESH);
while((my $key, my $value) = each (%namehash))
    {
    print OUT "\nTERM LINEAGE FOR " . uc($key) . "\n";
    my @valuelist = split(/ /,$value);
    my @cumlist;
    my %marked;
    foreach $meshno (@valuelist)
      {
      push(@cumlist, $meshno);
      while ($meshno =~ /\.[0-9]+$/)
          {
          $meshno = $`;
          push(@cumlist, $meshno);
          }
      }
    @cumlist = grep { $marked{$_}++; $marked{$_} == 1; }@cumlist;
```

```
    @cumlist = reverse(sort(@cumlist));
    foreach my $thing (@cumlist)
        {
        print OUT "$thing $numberhash{$thing}\n";
        }
    }
exit;
```

Python Script

```python
#!/usr/local/bin/python
import re
import string
mesh = open('C:\\big\\d2009.bin', "r")
out = open("mesh.out", "w")
namehash = {}
numberhash = {}
for line in mesh:
  namematch = re.search(r'MH = (.+)$', line)
  if (namematch):
    name = namematch.group(1)
  numbermatch = re.search(r'MN = (.+)$',line)
  if numbermatch:
      number = numbermatch.group(1)
      numberhash[str(number)] = name
      if namehash.has_key(name):
          namehash[name] = namehash[name] + ' ' + str(number)
      else:
          namehash[name] = str(number)
mesh.close()
keylist = namehash.keys()
keylist.sort()
for key in keylist:
   print>>out,"\nTERM LINEAGE FOR ", key.upper()
   cumlist = []
   item_list = namehash[key].split(" ")
   for phrase in item_list:
       cumlist.append(phrase)
       while re.search(r'\.', phrase):
         phrase = re.sub(r'\.[0-9]+$',"", phrase)
         cumlist.append(phrase)
   U = []
   for item in cumlist:
       if item not in U:
           U.append(item)
   U.sort()
   for thing in U:
       print>>out, thing, numberhash[str(thing)]
exit
```

Ruby Script

```
#!/usr/local/bin/ruby
mesh = File.open("c\:\\big\\d2009.bin", "r")
out = open("mesh.out", "w")
namehash = Hash.new("")
numberhash = Hash.new("")
name = ""
number = ""
mesh.each_line do
  |line|
  if (line =~ /MH \= (.+)$/)
    name = $1
  end
  if (line =~ /MN \= (.+)$/)
     number = $1
     numberhash[number] = name
     if namehash.has_key?(name)
       namehash[name] = namehash[name] + " " + number
     else
       namehash[name] = number
     end
   end
end
mesh.close
namehash.keys.sort.each do
  |k|
  out.puts "\nTERM LINEAGE FOR " + k
  cumlist = []
  item_list = namehash[k].split(" ")
  item_list.each do
    |phrase|
    cumlist.push(phrase)
    while (phrase =~ /\.[0-9]+$/)
       phrase = $`
       cumlist.push(phrase)
    end
  end
  cumlist.sort.uniq.each {|it| out.puts it + " " + numberhash[it]}
end
exit
```

5.1.2 Analysis

The output file exceeds 8 megabytes (MB) in length. Here are a few output terms, and their MeSH lineages:

Term lineage for retinoschisis
C11.768.585.865 Retinoschisis
C11.768.585 Retinal Degeneration
C11.768 Retinal Diseases
C11 Eye Diseases

Term lineage for core binding factor beta subunit
D12.776.930.155.400 Core Binding Factor Beta Subunit
D12.776.930.155 Core Binding Factors
D12.776.930 Transcription Factors
D12.776 Proteins
D12 Amino Acids, Peptides, and Proteins

Term lineage for giant cells, foreign-body
A11.118.637 Leukocytes
A15.145.229.637 Leukocytes
A15.382.490 Leukocytes
A11.118.637.555 Leukocytes, Mononuclear
A15.145.229.637.555 Leukocytes, Mononuclear
A15.382.490.555 Leukocytes, Mononuclear
A15.378 Hematopoietic System
A11.148 Bone Marrow Cells
A15.378.316 Bone Marrow Cells
A12.207.152 Blood
A15.145 Blood
A11.118 Blood Cells
A15.145.229 Blood Cells
A11.329.372.376 Giant Cells, Foreign-Body
A11.502.376 Giant Cells, Foreign-Body
A11.627.624.480.376 Giant Cells, Foreign-Body
A11.733.397.376 Giant Cells, Foreign-Body
A15.382.680.397.376 Giant Cells, Foreign-Body
A15.382.812.522.376 Giant Cells, Foreign-Body
A11.329 Connective Tissue Cells
A11 Cells
A11.502 Giant Cells
A11.118.637.555.652 Monocytes
A11.148.580 Monocytes
A11.627.624 Monocytes
A11.733.547 Monocytes
A15.145.229.637.555.652 Monocytes

A15.378.316.580 Monocytes
A15.382.490.555.652 Monocytes
A15.382.680.547 Monocytes
A15.382.812.547 Monocytes
A11.627 Myeloid Cells
A11.733 Phagocytes
A15.382.680 Phagocytes
A15.382 Immune System
A15 Hemic and Immune Systems
A11.329.372 Macrophages
A11.627.624.480 Macrophages
A11.733.397 Macrophages
A15.382.680.397 Macrophages
A15.382.812.522 Macrophages
A15.382.812 Reticuloendothelial System
A12.207 Body Fluids
A12 Fluids and Secretions

When we examine the multilineage ancestry of "foreign body giant cells", we see that MeSH is not a tree hierarchy. This means that the MeSH data structure is highly complex and requires some computational know-how to fully explore all the term relationships.

5.2 Creating a MeSH Database

The primary feature that distinguishes a database from a data object, such as a string variable or a list or a dictionary, is persistence. When you exit a database application, the data and the data structures created for the database, all persist, somewhere, on a hard drive. When you start the database at a later time, it is not necessary to port the data back into the application or to rebuild tables and relational structures; everything is waiting for you. A data object, created by a script, drops out of existence when the script stops executing. Data objects need to be rebuilt with each new execution of a script.

Perl, Python, and Ruby all have access to external database modules that can build database objects that exist as external files, separate from the script and from the scripting language; they persist after the script has executed. These database objects can be called from any script, with the contained data accessed quickly, with a simple command syntax.

5.2.1 Script Algorithm

1. Call the external database modules.
2. Using the required syntax for the chosen language, name and create a new database object and tie the object to a dictionary (hash) object.

3. Create a dictionary object and assign key–value pairs corresponding to the codes and the terms of MeSH records (steps 4–9).

4. Open the ASCII version of the MeSH file, for reading. The download site is

   http://www.nlm.nih.gov/mesh/filelist.html

   Download the d2009.bin file (referred to as the ASCII MeSH download file). This plain-text file is about 28 MB in length and contains over 25,000 MeSH records. The record format of the d2009.bin file is described in the appendix.

5. Open a file for writing. This file will receive the full hierarchy of the MeSH terms.

6. Parse through the MeSH file, line by line.

7. When a line that begins with "MH = " occurs, capture the MeSH term found on the line.

8. When a line that begins with "MN = " occurs, capture the MeSH number found on the line.

9. Because the MeSH term always occurs before the MeSH number, the MeSH number will always be the number that corresponds with the previously captured MeSH term. Use the captured number and the captured term to create a new key–value pair for a dictionary object.

10. Close the database object.

Perl Script

```perl
#!/usr/local/bin/perl
use Fcntl;
use SDBM_File;
tie %mesh_hash, "SDBM_File", 'mesh', O_RDWR|O_CREAT|O_EXCL, 0644;
open (TEXT, "c\:\\big\\d2009.bin")||die"Can't open file";
$line = " ";
while ($line ne "")
    {
    $line = <TEXT>;
    if ($line =~ /^MH = (.+)$/)
          {
          $term = $1;
          }
    if ($line =~ /^MN = (.+)$/)
          {
          $number = $1;
          $mesh_hash{$number} = $term;
          }
    }
untie %mesh_hash;
exit;
```

Python Script

```python
#!/usr/local/bin/python
import anydbm, string, re
mesh_hash = anydbm.open('mesh', 'n')
mesh_file = open('C:\\big\\d2009.bin', "r")
for line in mesh_file:
  namematch = re.search(r'^MH = (.+)$', line)
  if (namematch):
    name = namematch.group(1)
  numbermatch = re.search(r'^MN = (.+)$',line)
  if numbermatch:
    number = numbermatch.group(1)
    mesh_hash[str(number)] = name
mesh_hash.close()
exit
```

Ruby Script

```ruby
#!/usr/local/bin/ruby
require 'dbm'
mesh_file = File.open('c:/big/d2009.bin', 'r')
db = DBM.open('mesh')
term = " "
mesh_file.each_line do
  |line|
  if (line =~ /^MH = (.+)$/)
    term = $1.to_s
  end
  if (line =~ /^MN = (.+)$/)
    number = $1.to_s
    db[number] = term
  end
end
db.close
exit
```

5.2.2 Analysis

The created database exists as an external file. The name of the prefix to the external file is a parameter provided in the script statement that creates the database object: "mesh" in this example.

5.3 Reading the MeSH Database

Once a database has been created, the data can be efficiently called from within the same script that created the database, or from any other script, at any time. You only need to remember two things: (1) not to delete the created database file, and (2) not to assume that the database is an unchanged object; other scripts can add to or modify the contents of the original database object.

5.3.1 Script Algorithm

1. Call the external database modules for your chosen language.
2. Open the external database and tie the database to a dictionary object. The example script requires the existence of an external database file ("mesh") that was created in the previous section of this chapter.
3. Read and print every key–value pair in the dictionary.
4. Untie the database from the dictionary object.

Perl Script

```
#!/usr/local/bin/perl
use Fcntl;
use SDBM_File;
tie %mesh_hash, "SDBM_File", 'mesh', O_RDWR, 0644;
while(($key, $value) = each (%mesh_hash))
   {
   print "$key => $value\n";
   }
untie %mesh_hash;
exit;
```

Python Script

```
#!/usr/local/bin/python
import anydbm, string, re
mesh_hash = anydbm.open('mesh')
for number in mesh_hash.keys():
   print number, mesh_hash[number]
mesh_hash.close()
exit
```

Ruby Script

```
#!/usr/local/bin/ruby
require 'dbm'
db = DBM.open('mesh')
db.each {|number,term| puts number + " hello " + term}
db.close
exit
```

5.3.2 Analysis

External databases achieve data persistence, but there is a cost:

1. You must keep track of the file names and directory locations of your external databases. If you're not careful, you can create hundreds of database files sprinkled throughout your file system, all with the same filename.
2. Programmers sometimes forget that external databases, unlike local variables, carry their data forever. You may have inadvertently used an external database file, with a secondary script, adding to its contents, without anticipating the unintended consequences that occur when another script accesses the same database file.
3. External database files may return data slower than data objects created within a script. In general, you will find that key–value pairs that are accessed repeatedly, within a script, will be accessed more quickly than seldom-accessed pairs. Built-in memorization subroutines, whereby repeatedly accessed variables are stored and retrieved from RAM memory, mitigate slow access times required for hard-disk retrievals.

5.4 Creating an SQLite Database for MeSH

SQL (Systems Query Language, pronounced like the word "sequel") is a specialized language used to query relational databases. SQL allows programmers to connect with large, complex server-based network databases. Learning SQL is like learning any programming language. A high level of expertise is needed to install and implement software that creates server-based relational databases and that responds to multiuser client-based SQL queries. Fortunately, users of Perl, Ruby, and Python all have easy access to SQLite. SQLite is a no-cost, open source program that you can use to build a relational database on your own computer. The SQLite database will respond to SQL statements appearing within your Perl, Python, or Ruby scripts.

Database statements perform a small collection of tasks: create a new database file, create a new table within the file, make records for the table, delete records, modify records, and select records or parts of records based on a query's selection criteria.

In the prior two sections, we showed how you could load a large text corpus into a database object in Perl, Python, or Ruby. In this section, we will show how you can perform the same task with SQLite, yielding a relational database file, residing on your own computer, that can be modified or queried at a later time, with other scripts.

First, you must install SQLite and the language-specific interface to SQLite for your preferred programming language. This can be easily accomplished, and the instructions for acquiring and installing SQLite are described in the appendix. Once done, you can access the MeSH ASCII data set, exactly as we did in the prior two sections. Then use Perl, Python, or Ruby scripts to load the terms, and their MeSH codes, into a relational database table.

5.4.1 Script Algorithm

1. Call the SQL database module into your script.
2. Create a dictionary object and assign key–value pairs corresponding to the codes and the terms of MeSH records (steps 3–8).
3. Open the ASCII version of the MeSH file, for reading.
   The download site is:

   http://www.nlm.nih.gov/mesh/filelist.html

   Download the d2009.bin file (referred to as the ASCII MeSH download file). This plain-text file is about 28 MB in length and contains over 25,000 MeSH records. The record format of the d2009.bin file is described in the appendix.
4. Open a file for writing. This file will receive the full hierarchy of the MeSH terms.
5. Parse through the MeSH file, line by line.
6. When a line that begins with "MH = " occurs, capture the MeSH term found on the line.
7. When a line that begins with "MN = " occurs, capture the MeSH number found on the line.
8. Because the MeSH term always occurs before the MeSH number, the MeSH number will always be the number that corresponds with the previously captured MeSH term. Use the captured number and the captured term to create a new key–value pair for a dictionary object.
9. Create a table for the new database object. We will call the new table "mesh".
10. Specify that the new table ("mesh") will contain rows occupied by two values. Each value in the row will be a character variable.
11. Execute the SQL statements that prepare the table.
12. Prepare the table for an SQL transaction in which it will receive pairs of values to be inserted as records for the table.
13. Parse through the dictionary object (prepared in steps 3–8), that contains MeSH codes as keys, and MeSH terms as the corresponding values to the dictionary keys. As each pair of MeSH code and MeSH term is parsed, insert them as records for the "mesh" table of your newly created SQLite database.
14. When the key–value pairs of the dictionary object have been parsed, close the INSERT transaction with the COMMIT statement, thus populating the "mesh" table within the database object.

Perl Script

```
#!/usr/local/bin/perl
use DBI;
open (TEXT, "c\:\\big\\d2009.bin")||die"Can't open file";
$line = " ";
while ($line ne "")
```

```perl
      {
      $line = <TEXT>;
      if ($line =~ /^MH = (.+)$/)
            {
            $term = $1;
            }
      if ($line =~ /^MN = (.+)$/)
            {
            $number = $1;
            $mesh_hash{$number} = $term;
            }
      }
close TEXT;
my $dbh = DBI->connect("dbi:SQLite:dbname=dbfile","","");
my $sth = $dbh->prepare("CREATE TABLE mesh (number VARCHAR(64),
term VARCHAR(64))");
$sth->execute;
$sth = $dbh->prepare("INSERT INTO mesh (number,term) VALUES(?,?)");
$dbh->do( "BEGIN TRANSACTION");
while ((my $key, my $value) = each(%mesh_hash))
     {
     $sth->execute( $key, $value );
     }
$dbh->do( "COMMIT" );
exit;
```

The resulting database is an external file, in the same directory as your Perl script, named "dbfile".

Python Script

```python
#!/usr/local/bin/python
from pysqlite2 import dbapi2 as sqlite
import string, re, os
mesh_file = open('C:\\big\\d2009.bin', "r")
mesh_hash = {}
entry = ()
for line in mesh_file:
     namematch = re.search(r'^MH = (.+)$', line)
     if (namematch):
     name = namematch.group(1)
     numbermatch = re.search(r'^MN = (.+)$',line)
     if numbermatch:
     number = numbermatch.group(1)
     mesh_hash[str(number)] = name
con=sqlite.connect('test1.db')
cur=con.cursor()
cur.executescript("""
```

```
        create table mesh
        (
        name varchar(64),
        term varchar(64)
        );
        """)
for key, value in mesh_hash.iteritems():
        entry = (key, value)
        cur.execute("insert into mesh (name, term) values (?, ?)",
entry)
con.commit()
exit
```

The resulting database is an external file, in the same directory as your Python script, named "test1.db".

Ruby Script

```
#!/usr/local/bin/ruby
require 'sqlite3'
db = SQLite3::Database.new( "test.db" )
mesh_file = File.open('c:/big/d2009.bin', 'r')
db_hash = Hash.new()
term = " "
mesh_file.each_line do
  |line|
  if (line =~ /^MH = (.+)$/)
    term = $1.to_s
  end
  if (line =~ /^MN = (.+)$/)
    number = $1.to_s
    db_hash[number] = term
  end
end
sql = <<SQL
    create table mesh (
      a varchar2(64),
      b varchar2(64)
    );
SQL
db.execute_batch( sql )
db.transaction
db_hash.each {|k,v| db.execute("insert into mesh values (?,?)",
k,v)}
db.commit
exit
```

The resulting database is an external file, in the same directory as your Ruby script, named "test.db".

5.4.2 Analysis

Many database programmers write programs that connect to an existing database, often residing on a remote server, replying to complex queries on the contained data. Some database programmers concentrate on writing programs that can add single reports to an existing database, entered by multiple users at multiple sites, a service that a hospital information systems might provide. Creating a new relational database by porting data from a large biomedical data set is a task more suited to a biomedical scientist than to a database programmer. The typical SQL data insertion statement commits the insertion and waits for the data to be loaded to disk before preparing the next insertion statement. This step, repeated thousands or millions of time, greatly impedes scripts such as ours, which loaded a data set into a database. In our script, we used a method that opens a transaction process that parses our entire data set before committing the process. This simple trick permits the rapid execution of the script. On my modest computer (2.5 GHz CPU, with 512 MB of RAM memory), it takes about 15 seconds to build the entire database, from the approximately 28 MB MeSH file.

5.5 Reading the SQLite MeSH Database

Once you have created an SQL relational database, as we have done in the prior section, you can access the data through a Perl, Python, or Ruby interface. SQL provides several ways of fetching and organizing data from a database, and there are a great many books written on the subject. Most database programmers settle into a tried-and-true set of SQL statements that suit their recurring needs. For now, we will write a very simple script that connects to the database created in the prior section, enters the only table that we prepared, and fetches all of the data elements. Using terminology from the field of relational databases, this corresponds to the rows and the column entries for the table.

5.5.1 Script Algorithm

1. Call the SQL module.
2. Open a file for writing. This file will receive the data elements extracted from the database table.
3. Connect to the SQLite database.
4. In the prior section, we created a table named "mesh" and populated every row of the table with two elements, corresponding to a MeSH code and its corresponding MeSH term. Using the "select" statement, select every data element from every 2-element row of "mesh" table.
5. Execute the SQL select statement, printing the contents of each successive row into the file prepared in step 2.
6. Exit.

Perl Script

```perl
#!/usr/local/bin/perl
use DBI;
open(OUT, ">meshdb.txt");
my $dbh = DBI->connect("dbi:SQLite:dbname=dbfile","","");
$sth = $dbh->prepare("SELECT number, term FROM mesh");
$sth->execute;
while (@row = $sth->fetchrow_array())
  {
  print OUT "@row\n";
  }
exit;
```

Python Script

```python
#!/usr/local/bin/python
from pysqlite2 import dbapi2 as sqlite
out_text = open("meshdb.txt", "w")
con=sqlite.connect('test1.db')
cur=con.cursor()
cur.execute("select * from mesh")
for row in cur:
  print>>out_text, row[0], row[1]
exit
```

Ruby Script

```ruby
#!/usr/local/bin/ruby
require 'sqlite3'
fout = File.open("meshdb.txt", "w")
db = SQLite3::Database.new( "test.db" )
db.execute("select * from mesh") do
 |row|
 fout.puts row[0] + " " + row[1]
end
```

5.5.2 Analysis

The output file of the script, meshdb.txt, is approximately 2 MB in length and contains about 47,000 codes and corresponding terms. Here are a few lines of the output file:

D02.455.526.728.468 Mustard Gas
I01.880.735.580 Needle Sharing
G07.700.320.500.325.180 Embryonic Development
F04.754.720.864.363 Countertransference (Psychology)
C02.800.801.220 Condylomata Acuminata
C04.651.600 Multiple Endocrine Neoplasia

On my modest computer (2.5 GHz CPU, with 512 MB of RAM memory), it takes about 1 second to download the entire database.

In the script, we used the SQL "select" statement to extract data from rows. The "select" statement, along with about half a dozen optional parameters, is the key method used by professional programmers to interrogate relational databases and organize the extracted data. Healthcare professionals who master the intricacies of SQL's "select" statement will find that they can perform a wide range of database tasks with ease.

Exercises

1. In Perl, Python, or Ruby, determine the total number of different terms in MeSH.
2. In Perl, Python, or Ruby, write a script that parses through a text file and creates lists of every single word term in the file, every two word term in the file, and every three word term in the file.
3. In Perl, Python, or Ruby, write a script that lists every term in MeSH that is classified as a disease (i.e., that has "disease" as an ancestor term).
4. In Perl, Python, or Ruby, write a script that lists the immediate parent class of each term in MeSH.
5. When we prepared the SDBM database and the relational database from the MeSH file, we overlooked a curious property of MeSH terms. A single MeSH term may have more than one MeSH code. For each MeSH record, our script assigns a single MeSH code to each MeSH term. In Perl, Python, or Ruby, modify the script that uses SDBM to make a database object, or the script that uses SQLite to create a MeSH database, so that every MeSH code that corresponds to a MeSH term will be included as a database record.
6. In Perl, Python, or Ruby, modify the script that reads the SQLite database, so that the output file is sorted alphabetically, by MeSH term.
   Hint: Use the "select" method's optional "order by" parameter.

6

THE INTERNATIONAL CLASSIFICATION OF DISEASES

The International Classification of Diseases (ICD) is a nomenclature of the diseases occurring in humans, with each listed disease assigned a unique identifying code. The ICD is owned by the World Health Organization, but can be used freely by the public. The currently used version of ICD is version 10 (ICD10). The World Health Organization also produces a specialized cancer nomenclature, known as the ICD-O (ICD-Oncology). The ICD is used worldwide. In the United States, ICD codes are used by the CDC (Centers for Disease Control and Prevention) to designate the causes of death listed on death certificates.

Causes of death listed in the CDC mortality record are represented by ICD10 codes. We will be using ICD and ICD-O in scripts throughout this book.

6.1 Creating the ICD Dictionary

If we have a computer computer-parsable list of ICD codes, we can write a short program that assigns human-readable terms (full names of diseases) to the codes in the mortality files.

An electronic version of the ICD is provided from the CDC, under the filename "cach10.txt". The each10.txt file is available by anonymous ftp from the ftp.cdc.gov Web server at:

/pub/Health_Statistics/NCHS/Publications/ICD10/each10.txt

Here are the first few lines of this file:

A00Cholera
A00.0Cholera due to Vibrio cholerae 01, biovar cholerae
A00.1Cholera due to Vibrio cholerae 01, biovar el tor
A00.9Cholera, unspecified
A01Typhoid and paratyphoid fevers
A01.0Typhoid fever
A01.1Paratyphoid fever A
A01.2Paratyphoid fever B
A01.3Paratyphoid fever C
A01.4Paratyphoid fever, unspecified

A02Other salmonella infections

A02.0Salmonella gastroenteritis

We will create a dictionary data object consisting of ICD codes (as dictionary keys) and their corresponding terms (as dictionary values).

6.1.1 Script Algorithm

1. Open the each10.txt file.
2. Put the entire file into a string variable.
3. Split the string variable wherever the newline character is followed by an ICD code.
4. For each split item, add the code (as the key) and the term (as the value) to the dictionary.
5. Print out all of the dictionary key–value pairs, with the keys sorted alphabetically, to the each10.out file.

Perl Script

You will need to place the each10.txt file in the same subdirectory as the Perl script.

```
#!/usr/local/bin/perl
open (ICD, "c\:\\ftp\\each10.txt")||die"cannot";
undef($/);
$line = <ICD>;
close ICD;
@linearray = split(/\n(?=[ ]*[A-Z][0-9\.]{1,5})/, $line);
foreach $thing (@linearray)
   {
   if ($thing =~ /^ *([A-Z][0-9\.]{1,5}) ?/)
      {
      $code = $1;
      $term = $';
      $term =~ s/\n//;
      $term =~ s/[ ]+$//;
      $dictionary{$code} = $term;
      }
   }
open (TEXT, ">c\:\\ftp\\each10.out");
foreach $key (sort keys %dictionary)
   {
   printf TEXT ("%-8.08s %s\n", $key, $dictionary{$key});
   }
close TEXT;
exit;
```

Python Script

```
#!/usr/local/bin/python
import sys, os, re, string
linearray = []
dictionary = {}
code = ""
term = ""
in_text = open('C:\\ftp\\each10.txt', "r")
in_text_string = in_text.read()
in_text.close()
linearray = in_text_string.split("\n")
for item in linearray:
    m = re.search(r'^[ \*]*([A-Z][0-9\.]{1,7}) ?([^0-9].+)', item)
    if m:
        code = m.group(1)
        term = m.group(2)
        dictionary[code] = term
out_text = open("c:\\ftp\\each10.out", "w")
dict_list = dictionary.keys()
sort_list = sorted(dict_list)
for i in sort_list:
  print>>out_text, "%-8.08s %s" % (i, dictionary[i])
out_text.close()
exit
```

Ruby Script

```
#!/usr/local/bin/ruby
f = File.open("c:/ftp/each10.txt")
dictionary = Hash.new("")
f.each do
  |line|
  next unless (line =~ /^[ \*]*([A-Z][0-9\.]{1,7}) ?([^0-9].+)/)
  code = $1
  term = $2
  dictionary[code] = term
end
f.close
fout = File.open("c:/ftp/each10.out", "w")
dictionary.keys.sort.each {|k| fout.printf "%-8.08s %s\n", k,
dictionary[k]}
exit
```

6.1.2 Analysis

The output file, each10.out, contains about 9,270 code–term pairs and has a length of about 440,000 bytes.

```
A00         Cholera
A00.0       Cholera due to Vibrio cholerae 01, biovar cholerae
A00.1       Cholera due to Vibrio cholerae 01, biovar el tor
A00.9       Cholera, unspecified
A01         Typhoid and paratyphoid fevers
A01.0       Typhoid fever
A01.1       Paratyphoid fever A
A01.2       Paratyphoid fever B
A01.3       Paratyphoid fever C
A01.4       Paratyphoid fever, unspecified
A02         Other salmonella infections
A02.0       Salmonella gastroenteritis
A02.1       Salmonella septicemia
A02.2       Localized salmonella infections
A02.8       Other specified salmonella infections
A02.9       Salmonella infection, unspecified
```

6.2 Building the ICD-O (Oncology) Dictionary

The ICD-Oncology (International Classification of Diseases, Oncology) is a specialized vocabulary created by the World Health Organization. ICD-O contains the dictionary of neoplasm codes and terms used by cancer registrars. The U.S. National Cancer Institute has been collecting millions of deidentified cancer records in its SEER (Surveillance Epidemiology and End Results) program. The SEER data records represent the names of neoplasms by their ICD-O codes. We will be using the ICD-O nomenclature in several different projects.

The ICD-O (Oncology) contains codes for 9,769 neoplasm terms, and is freely available from SEER, as a PDF file at

http://seer.cancer.gov/icd-o-3/sitetype.icdo3.d08152007.pdf

The SEER file, reduced to ASCII text, is available at

http://www.julesberman.info/book/icdo3.txt

Additional information on the ICD-O file is found in the appendix.
Here are a few lines from the icdo3.txt file:

8000/3 Neoplasm, malignant
8001/3 Tumor cells, malignant
8002/3 Malignant tumor, small cell type
8003/3 Malignant tumor, giant cell type
8004/3 Malignant tumor, spindle cell type
8005/3 Malignant tumor, clear cell type
CARCINOMA, NOS 801 8010/2 Carcinoma in situ, NOS
8010/3 Carcinoma, NOS

8011/3 Epithelioma, malignant

8012/3 Large cell carcinoma, NOS

8013/3 Large cell neuroendocrine carcinoma

8014/3 Large cell carcinoma with rhabdoid phenotype

8015/3 Glassy cell carcinoma

CARCINOMA, UNDIFF., NOS 802 8020/3 Carcinoma, undifferentiated type, NOS

8021/3 Carcinoma, anaplastic type, NOS

8022/3 Pleomorphic carcinoma

The ICD-O file can be parsed into code–term pairs.

6.2.1 Script Algorithm

1. Open the icdo3.txt file.
2. Parse the icdo3.txt file, line by line.
3. Each line begins with a code, consisting of four digits followed by a slash, followed by one digit, followed by a space, followed by the term. Create a regex expression for the line, placing the five digits from the code into a key variable, and the term into a value variable, for a hash object.
4. Sort the keys of the hash object, and print the key (code)–value (term) pairs.

Perl Script

```perl
#!/usr/local/bin/perl
open (ICD, "c\:\\ftp\\icdo3\.txt");
$line = " ";
while ($line ne "")
  {
  $line = <ICD>;
  if ($line =~ /([0-9]{4})\/([0-9]{1}) +/o)
    {
    $code = $1 . $2;
    $term = $';
    $term =~ s/ *\n//o;
    $term = lc($term);
    $dictionary{$code} = $term;
    }
  }
close ICD;
foreach $icd_code (sort(keys(%dictionary)))
  {
  print $icd_code . " " . $dictionary{$icd_code} . "\n";
  }
exit;
```

Python Script

```
#!/usr/local/bin/python
import sys, os, re, string
f = open("c:\\ftp\\icdo3.txt", "r")
codehash = {}
for line in f:
  linematch = re.search(r'([0-9]{4})\/([0-9]{1}) +(.+)$', line)
  if (linematch):
    icdcode = linematch.group(1) + linematch.group(2)
    term = string.rstrip(linematch.group(3))
    codehash[icdcode] = term
f.close
keylist = codehash.keys()
keylist.sort()
for item in keylist:
  print item, codehash[item]
exit
```

Ruby Script

```
#!/usr/local/bin/ruby
f = File.open("c:/ftp/icdo3.txt")
fout = File.open("SEER.OUT", "w")
codehash = Hash.new("")
f.each do
  |line|
  next unless (line =~ /([0-9]{4})\/([0-9]{1}) +/)
  icdcode = $1 << $2
  term = $'.chomp!
  codehash[icdcode] = term
end
f.close
codehash.keys.sort.each {|key| puts "#{key} #{codehash[key]}"}
exit
```

6.2.2 Analysis

Here are a few of the code–term pairs from ICD-O:

99403 Hairy cell leukemia
99453 Chronic myelomonocytic leukemia, NOS
99463 Juvenile myelomonocytic leukemia
99483 Aggressive NK-cell leukemia
99503 Polycythemia vera
99603 Chronic myeloproliferative disease, NOS
99613 Myelosclerosis with myeloid metaplasia
99623 Essential thrombocythemia

99633 Chronic neutrophilic leukemia

99643 Hypereosinophilic syndrome

99803 Refractory anemia

99823 Refractory anemia with sideroblasts

99833 Refractory anemia with excess blasts

99843 Refract. anemia with excess blasts in transformation

99853 Refractory cytopenia with multilineage dysplasia

99863 Myelodysplastic syndr. with 5q deletion syndrome

99873 Therapy-related myelodysplastic syndrome, NOS

99893 Myelodysplastic syndrome, NOS

We will be using the ICD-Oncology dictionary in later chapters.

Exercises

1. The ICD10 nomenclature is composed of line records, with each line consisting of a code followed by a term. Using Perl, Python, or Ruby, write a script that reverses the order for each line entry, with the term preceding the code, and print the list of term/code pairs in alphabetical order.
2. Repeat Exercise 1, using ICD-O (instead of ICD10).
3. When we created the ICD Oncology dictionary, our output file listed each code (in ascending numeric order), followed by the term. Using your favorite programming language, revise the script to output each term (in alphabetical order) followed by the code number. Format the output so that each term's code number lines up in a column with all of the other term–code pairs.
4. The ICD10 is an imperfect nomenclature. Here are some of the entries:

    M95.2 Other acquired deformity of head

    M95.8 Other specified acquired deformities of musculoskeletal system

    M99.8 Other biomechanical lesions

    N00.8 Other

    N01.8 Other

    N02.8 Other

    N03.8 Other

    N04.8 Other

    N05.8 Other

    N06.8 Other

    N07.8 Other

    N11.8 Other chronic tubulo-interstitial nephritis

    N13.3 Other and unspecified hydronephrosis

    N13.8 Other obstructive and reflux uropathy

    Though every code in the nomenclature is unique, some of the terms are nonunique. There are many codes whose term is simply "Other". This oversight is a throwback to the time when nomenclatures were not designed to be computer-parsable. Human readers could look at a list of related terms and infer the intended meaning of the nonunique terms.

Using Perl, Python, or Ruby, write a script that parses the each10.out file (the dictionary of ICD10 codes and terms) and that collects all of the code–term pairs containing a nonunique term. Confer uniqueness on the nonunique terms by appending the parent term to the child term.

For example, there are numerous codes that have the same term, "Other". The term associated with N00.8 is "Other". We know that the parent term for N00.8 is N00, "Acute nephritic syndrome".

N00 Acute nephritic syndrome

N00.0 Minor glomerular abnormality

N00.1 Focal and segmental glomerular lesions

N00.2 Diffuse membranous glomerulonephritis

N00.3 Diffuse mesangial proliferative glomerulonephritis

N00.4 Diffuse endocapillary proliferative glomerulonephritis

N00.5 Diffuse mesangiocapillary glomerulonephritis

N00.6 Dense deposit disease

N00.7 Diffuse crescentic glomerulonephritis

N00.8 Other

N00.9 Unspecified

We can put the parent term as the addendum to the child term, "Other acute nephritic syndrome". This creates a new term that is unique from every other term in the dictionary and that conveys the full, intended meaning of N00.8.

After collecting the nonunique terms in the ICD10 nomenclature, your script should append the parent term to each of the nonunique terms, returning the new, unique term to the dictionary.

7
SEER

The Cancer Surveillance, Epidemiology, and End Results Program

SEER is the U.S. National Cancer Institute's Surveillance, Epidemiology, and End Results program. It is an amazing resource for information about the cancers that occur in the United States. One of the products of SEER is the Public Use data sets, which contain deidentified records on over 3.7 million cancers that have occurred between 1973 and 2006.

When you have over 3.7 million cancer cases to study, you can draw certain types of inferences that could not possibly be made with the data accumulated at any single medical institution.

The SEER site allows users to make data queries directly. If you would like to search the SEER data with the SEER search engine, the Web address is

http://seer.cancer.gov/canques/index.html

At the SEER site, users cannot make global queries (queries that compare every tumor in the database against every other tumor in the database, by every tumor type, and all at once). Global queries are what data mining is all about. Serious data miners write their own scripts to parse through the SEER public data files.

7.1 Parsing the SEER Data Files

Each SEER record is a cancer case, described by a series of 258 alphanumeric characters, in byte-assigned positions, described by a data dictionary document. When you have the byte locations for the data dictionary entries, you can easily write a short script that can extract and compile data any way you wish.

The appendix contains detailed instructions for downloading the SEER data files.

7.1.1 Script Algorithm

1. Open the directory containing the SEER data files. The files comprising the set of SEER data files are listed Exercise 1 of this chapter. I store my files in the c:\seer subdirectory, and, in this example script, this is the subdirectory that is opened.

2. Create a new dictionary in which the keys will be the age of the patient in the record, and the values will be the number of occurrences of the age in the SEER files.

3. Parse through each SEER data file, one line at a time.

4. For each SEER record, for every SEER file, extract the sequence of three consecutive bytes, corresponding to byte 25, 26, and 27 from the record. These bytes contain the age of the patient.

5. As each age is encountered, increment, by 1, the number of occurrences of the age (in the dictionary object that you have created in step 2).

6. After every SEER file has been parsed, print out the sorted list of ages and occurrences from the dictionary object.

Perl Script

```perl
#!/usr/local/bin/perl
opendir(SEERDIR, "c\:\\seer") || die ("Unable to open directory");
@files = readdir(SEERDIR);
closedir(SEERDIR);
chdir("c\:\\seer");
foreach $datafile (@files)
  {
  next if ($datafile !~ /\.txt/i);
  open (TEXT, $datafile);
  $line = " ";
  while ($line ne "")
    {
    $line = <TEXT>;
    $agehash{$&}++ if (substr($line,24,3) =~ /[01][0-9]{2}/)
    }
  }
foreach $age (sort(keys(%agehash)))
  {
  print "$age $agehash{$age}\n";
  }
exit;
```

Python Script

```python
#!/usr/local/bin/python
import sys, os, re
agehash = {}
filelist = os.listdir("c:\\seer")
os.chdir("c:\\seer")
for file in filelist:
  infile = open(file,'r')
  for line in infile:
    age = line[24:27]
    if re.search(r'[01][0-9]{2}',age):
```

```
        if agehash.has_key(age):
           agehash[age] = agehash[age] + 1
        else:
           agehash[age] = 1
  infile.close()
keylist = agehash.keys()
keylist.sort()
for item in keylist:
  print item, agehash[item]
exit
```

Ruby Script

```
#!/usr/local/bin/ruby
filelist = Dir.glob("c:/seer/*.TXT")
agehash = {}
filelist.each do
  |filepathname|
  seer_file = File.open(filepathname)
  seer_file.each do
    |line|
    if (line.slice(24,3) =~ /[01][0-9]{2}/)
      if agehash.has_key? $&
        agehash[$&] = agehash[$&] + 1
      else
        agehash[$&] - 1
      end
    end
  end
end
agehash.keys.sort.each {|key| puts "#{key} #{agehash[key]}"}
exit
```

7.1.2 Analysis

In the SEER files, age is a three-digit sequence that is occupies byte 25, 26, and 27 from each line record for each SEER file. Because Perl, Python, and Ruby count from zero (not from one), bytes 25, 26, and 27 correspond to the 24th, 25th, and 26th bytes of the line record.

The following columns depict the ages (from 000 to 115) and the number of SEER records for each age:

000 2656	006 1259	012 1377
001 2276	007 1189	013 1576
002 2421	008 1136	014 1838
003 2209	009 1073	015 2090
004 1810	010 1165	016 2384
005 1527	011 1209	017 2971

018 3753	050 45817	082 59696
019 4454	051 48099	083 54535
020 5576	052 50274	084 48517
021 6746	053 52985	085 42474
022 8097	054 55555	086 37190
023 9231	055 59398	087 31861
024 10676	056 62248	088 27004
025 11786	057 65734	089 22319
026 12714	058 68730	090 17470
027 13852	059 71403	091 14116
028 14306	060 75204	092 11171
029 15319	061 77693	093 8588
030 15807	062 80501	094 6372
031 16308	063 83004	095 4618
032 16597	064 85536	096 3331
033 17266	065 90965	097 2171
034 17724	066 92046	098 1539
035 18906	067 94336	099 1007
036 19256	068 95657	100 579
037 20003	069 97415	101 355
038 21032	070 97192	102 245
039 22136	071 97525	103 137
040 24763	072 97617	104 84
041 25583	073 97079	105 40
042 27472	074 94923	106 36
043 28812	075 92417	107 16
044 30537	076 89849	108 11
045 33185	077 85938	109 3
046 35222	078 81630	110 1
047 37600	079 76125	113 2
048 39872	080 70060	115 1
049 42087	081 65250	

A quick scan of the list shows that the mode (age with the greatest number of occurrences) is age 72, with 97,617 cancer occurrences. This confirms that cancer occurs most frequently among seniors.

7.2 Finding the Occurrences of All Cancers in the SEER Data Files

The SEER data files consist of individual records of individual cancer cases. We can write simple scripts that parse the entire data set, counting the occurrences of each type of cancer, and producing a ranked output.

Sex	220	24–24
Age at diagnosis	230	25–27
Year of Birth	240	28–31
Birth Place	250	32–34
Sequence Number--Central	380	35–36
Month of diagnosis	390	37–38
Year of diagnosis	390	39–42
Primary Site	400	43–46
Laterality	410	47–47
Histology (92-00) ICD-0-2	420	48–51
Behavior (92-00) ICD-0-2	430	52–52
Histologic Type ICD-0-3	522	53–56
Behavior Code ICD-0-3	523	57–57

Figure 7.1 Partial listing of the SEER data dictionary, providing the fields covered by bytes 24 to 57 of each data record.

7.2.1 Script Algorithm

1. Open the ICD-O nomenclature file and produce a dictionary object, with keys containing the five-digit code for each type of neoplasm, and the corresponding values consisting of the name of the neoplasm.
2. Read the directory containing the SEER files, collecting the name of each data file.
3. Change the directory to the path of the SEER files, so that your script can directly access the SEER files.
4. Set up a loop that will parse through every SEER file, line by line, so that, by the end of the loop, every record in SEER will be examined.
5. For each SEER record, extract bytes that contain the ICDO2 code and the ICDO3 code for each record (Figure 7.1).
6. If an ICDO3 code exists, let it override the ICDO2 code.
7. In a dictionary object, add the record's code as a key in the dictionary object, and increment the value of the object by 1 (i.e., increment the code's occurrence tally every time the code appears in a record).
8. After all of the SEER files are parsed, print out the key–value pairs from the created dictionary object (now containing the codes occurring in the SEER file, with their number of occurrences). For each code, provide the term-equivalent for the code (held in the nomenclature dictionary object), and display the output by sorting on the occurrence number.

Perl Script

```
#!/usr/local/bin/perl
open (ICD, "c\:\\ftp\\icdo3\.txt");
```

```perl
$line = " ";
while ($line ne "")
   {
   $line = <ICD>;
   if ($line =~ /([0-9]{4})\/([0-9]{1}) +/o)
      {
      $code = $1 . $2;
      $term = $';
      $term =~ s/ *\n//o;
      $term = lc($term);
      $dictionary{$code} = $term;
      }
   }
close ICD;
opendir(SEERDIR, "c\:\\seer") || die ("Unable to open directory");
@files = readdir(SEERDIR);
closedir(SEERDIR);
open (OUT, ">c\:\\ftp\\seerdist.txt");
chdir("c\:\\seer");
foreach $datafile (@files)
   {
   next if ($datafile !~ /.txt/i);
   open (TEXT, $datafile);
   $line = " ";
   while ($line ne "")
      {
      $line = <TEXT>;
      $dx = substr($line, 47, 5);
      $dx2 = substr($line, 52, 5);
      if (exists($dictionary{$dx2}))
         {
         $dx = $dx2;
         }
      $dxhash{$dx}++;
      }
   close TEXT;
   }
while (($key, $value) = each(%dxhash))
   {
   if (exists($dictionary{$key}))
      {
      $value = "0000000" . $value;
      $value = substr($value, -7, 7);
      push(@distarray, "$value $dictionary{$key}");
      }
   }
print join("\n", reverse(sort(@distarray)));
exit;
```

Python Script

```
#!/usr/local/bin/python
import sys, os, re, string
f = open("c:\\ftp\\icdo3.txt", "r")
fout = open("SEER.OUT", "w")
codehash = {}
subhash = {}
for line in f:
  linematch = re.search(r'([0-9]{4})\/([0-9]{1}) +(.+)$', line)
  if (linematch):
    icdcode = linematch.group(1) + linematch.group(2)
    term = string.rstrip(linematch.group(3))
    codehash[icdcode] = term
f.close()
filelist = os.listdir("c:\\seer")
os.chdir("c:\\seer")
for file in filelist:
  seer_file = open(file, "r")
  for line in seer_file:
    code1 = line[47:52]
    code2 = line[52:57]
    if codehash.has_key(code2):
      code1 = code2
      if subhash.has_key(code1):
        subhash[code1] = subhash[code1] + 1
      else:
        subhash[code1] = 1
keylist = subhash.keys()
for item in keylist:
  print>>fout, "%-7.7d %8s" % (subhash[item], codehash[item])
fout.close()
orderfile = open("c:\\ftp\\py\\SEER.OUT", "r")
line_array = orderfile.readlines()
line_array.sort()
line_array.reverse()
for item in line_array:
  print item,
exit
```

Ruby Script

```
#!/usr/local/bin/ruby
f = File.open("c:/ftp/icdo3.txt")
fout = File.open("SEER.OUT", "w")
codehash = Hash.new("")
subhash = Hash.new(0)
f.each do
  |line|
```

```
  next unless (line =~ /([0-9]{4})\/([0-9]{1}) +/)
  icdcode = $1 << $2
  term = $'.chomp!
  codehash[icdcode] = term
end
f.close
filelist = Dir.glob("c:/seer/*.TXT")
begin_time = Time.new.to_f
filelist.each do
  |filepathname|
  seer_file = File.open(filepathname)
  seer_file.each do
    |line|
    code1 = line.slice(47,5)
    code2 = line.slice(52,5)
    code1 = code2 if codehash.has_key? code2
    subhash[code1] = subhash[code1] + 1
  end
end
subhash.each do
  |key,value|
  if codehash.has_key?(key)
    fout.printf("%-7.07d %-s \n", value, codehash[key])
  end
end
fout.close
fout = File.open("SEER.OUT")
final = File.open("seer2.out","w")
final.puts(fout.readlines.sort.reverse.join)
puts "\nTime to parse SEER files - #{Time.new.to_f - begin_time}
seconds"
exit
```

7.2.2 Analysis

The output list consists of the names of 645 different types of neoplasms, and the number of their occurrences in the SEER data files, listed in decreasing order. Here are the first 20 items in the output list:

```
1021940 Adenocarcinoma, NOS
0333623 Infiltrating duct carcinoma, NOS
0247826 Squamous cell carcinoma, NOS
0182616 Carcinoma, NOS
0096537 Papillary trans. cell carcinoma
0063370 Squamous cell carcinoma in situ, NOS
0057571 Carcinoma in situ, NOS
0056558 Neoplasm, malignant
```

0050243 Transitional cell carcinoma, NOS

0050173 Small cell carcinoma, NOS

0048315 Malignant melanoma, NOS

0045778 Mucinous adenocarcinoma

0040929 Superficial spreading melanoma

0037543 Renal cell carcinoma

0037088 ML, large B-cell, diffuse

0036429 Multiple myeloma

0035630 Intraductal carcinoma, noninfiltrating, NOS

0034336 Lobular carcinoma, NOS

0030328 B-cell chr. lymph. leuk./small lymphocytic lymphoma

0029592 Large cell carcinoma, NOS

The frequencies of the different kinds of cancers has a Zipf distribution. The most frequently occurring form of cancer, adenocarcinoma, accounts for over 1 million cases. The 20th item on the list is large cell carcinoma, NOS (not otherwise specified). It accounts for about 30,000 cases. If we were to look at the frequencies of occurrence of cancers in the bottom half of the list, we would see that many of these cancers account for just a few dozen cancer cases. We can easily see that a few kinds of cancers account for the bulk of the cancers that occur in humans.

7.3 Finding the Age Distributions of the Cancers in the SEER Data Files

Diseases often occur in narrow age ranges within a population. When analyzing disease data, it is not always sufficient to know the average age of occurrence of a disease. You really need to know the distribution of disease occurrences over a range of human ages (usually 0 to 100 years). This is particularly true for diseases with a multimodal age distribution.

7.3.1 Script Algorithm

1. Create a dictionary object consisting of code–term pairs from the ICD nomenclature.
2. Open the directory containing the SEER public data files.
3. Put the list of files in the directory into an array.
4. Parse every line of every file in the array. Each line represents a SEER data record. There are over 3 million records that will be parsed by the script.
5. For each parsed line, extract the diagnosis (represented as a five-digit ICD code) and the age (represented by a three-digit number ranging from 000 to 115).
6. Bin each age into one of 20 bins by dividing the age by 5, taking the integer value of the result, and lumping all ages 95 and above to the same bin (the 20th bin).

7. For each record, increment (by 1) the number of occurrences of the diagnosis (for the record), in the bin corresponding to the patient age listed for the record, and put the record diagnosis and the incremented age distribution for the diagnosis, as a key–value pair for a dictionary (hash) object.

8. After all of the files are parsed, parse through the dictionary object containing all of the diagnostic code/age distributions (steps 4–7).

9. As each pair of diagnostic codes and age distributions is parsed, print the term corresponding to the diagnostic code (from the dictionary object produced in step 1), and the age distribution for the term on a separate line.

Perl Script

```perl
#!/usr/local/bin/perl
open (ICD, "c\:\\ftp\\icdo3\.txt");
$line = " ";
while ($line ne "")
  {
  $line = <ICD>;
  if ($line =~ /([0-9]{4})\/([0-9]{1}) +/o)
    {
    $code = $1 . $2;
    $term = $';
    $term =~ s/ *\n//o;
    $term = lc($term);
    $dictionary{$code} = $term;
    }
  }
close ICD;
opendir(SEERDIR, "c\:\\seer") || die ("Unable to open directory");
@files = readdir(SEERDIR);
closedir(SEERDIR);
open (OUT, ">c\:\\ftp\\seerdist.txt");
chdir("c\:\\seer");
foreach $datafile (@files)
  {
  next if ($datafile !~ /.txt/i);
  open (TEXT, $datafile);
  $line = " ";
  while ($line ne "")
    {
    $line = <TEXT>;
    $dx = substr($line, 47, 5);
    $dx2 = substr($line, 52, 5);
    if (exists($dictionary{$dx2}))
      {
      $dx = $dx2;
      }
```

```perl
    unless (exists($dxhash{$dx}))
      {
      $dxhash{$dx} = "0 0 0 0 0 0 0 0 0 0 0 0 0 0 0 0 0 0 0 0";
      }
    $age_at_dx = substr($line,24,3);
    if ($age_at_dx > 95)
       {
       $age_at_dx = 95;
       }
    $age_at_dx = int($age_at_dx / 5);
    @agearray = split(" ", $dxhash{$dx});
    $agearray[$age_at_dx]++;
    $dxhash{$dx} = join(" ", @agearray);
    }
  close TEXT;
  }
while (($key, $value) = each(%dxhash))
  {
  if (exists($dictionary{$key}))
     {
     push(@distarray,"$dictionary{$key}\|$value");
     }
  }
print OUT join("\n", sort(@distarray));
close OUT;
exit;
```

Python Script

```python
#!/usr/local/bin/python
import os, re, string
f - open("c:\\ftp\\icdo3.txt", "r")
fout = open("SEER.OUT", "w")
codchash = {}
subhash = {}
agearray = []
for line in f:
  linematch = re.search(r'([0-9]{4})\/([0-9]{1}) +(.+)$', line)
  if (linematch):
    icdcode = linematch.group(1) + linematch.group(2)
    term = string.rstrip(linematch.group(3))
    codehash[icdcode] = term
f.close()
filelist = os.listdir("c:\\seer")
os.chdir("c:\\seer")
for file in filelist:
  seer_file = open(file, "r")
  for line in seer_file:
    code1 = line[47:52]
    code2 = line[52:57]
```

```python
    if codehash.has_key(code2):
      code1 = code2
    age_at_dx = int(line[24:27])
    if (age_at_dx > 95):
      age_at_dx = 95
    age_at_dx = int(age_at_dx/5)
    if not subhash.has_key(code1):
      subhash[code1] = "0 0 0 0 0 0 0 0 0 0 0 0 0 0 0 0 0 0 0 0"
    agearray = subhash[code1].split(" ")
    old_value = agearray[age_at_dx]
    new_value = int(old_value) + 1
    agearray[age_at_dx] = str(new_value)
    subhash[code1] = " ".join(agearray)
keylist = subhash.keys()
for item in keylist:
  if codehash.has_key(item):
    print>>fout, codehash[item] + "|" + subhash[item]
fout.close()
orderfile = open("c:\\ftp\\py\\SEER.OUT", "r")
line_array = orderfile.readlines()
line_array.sort()
for item in line_array:
  print item,
exit
```

Ruby Script

```ruby
#!/usr/local/bin/ruby
f = File.open("c:/ftp/icdo3.txt")
dxhash = Hash.new("")
codehash = Hash.new("")
distarray = []
fout = File.open("SEER.OUT", "w")
f.each do
  |line|
  next unless (line =~ /([0-9]{4})\/([0-9]{1}) +/)
  icdcode = $1 << $2
  term = $'.chomp!
  codehash[icdcode] = term
end
f.close
filelist = Dir.glob("c:/seer/*.TXT")
filelist.each do
  |filepathname|
  seer_file = File.open(filepathname, "r")
  seer_file.each do
    |line|
    code1 = line.slice(47,5)
    code2 = line.slice(52,5)
    code1 = code2 if codehash.has_key? code2
```

```
    unless dxhash.has_key? code1
      dxhash[code1] = "0 0 0 0 0 0 0 0 0 0 0 0 0 0 0 0 0 0 0 0 0"
    end
    age_at_dx = line.slice(24,3)
    next if age_at_dx !~ /[01][0-9]{2}/
    age_at_dx = age_at_dx.to_f
    if (age_at_dx > 95)
      age_at_dx = 95
    end
    age_at_dx = (age_at_dx / 5).to_i
    agearray = dxhash[code1].split(" ")
    agearray[age_at_dx] = agearray[age_at_dx].to_i
    agearray[age_at_dx] = agearray[age_at_dx] + 1
    dxhash[code1] = agearray.join(" ")
  end
end
dxhash.each do
  |key, value|
  if codehash.has_key? key
    distarray.push("#{codehash[key]}\|#{value}")
  end
end
print distarray.sort.join("\n")
exit
```

7.3.2 *Analysis*

Here are 18 consecutive diagnoses and age distributions from the output file. Each diagnostic entity is followed by the number of records of the tumor, in the SEER data files, for each five-year age range from 0 years of age up to age 95 (and above).

Struma ovarii, malignant
0 0 0 0 1 1 2 1 3 4 1 1 0 0 1 1 0 0 0 0

Subcutaneous panniculitis-like T-cell lymphoma
0 0 0 2 0 0 0 3 1 0 3 5 2 4 2 2 2 0 0 0

Subependymoma
0 0 0 1 1 2 2 2 4 4 3 7 3 1 1 1 0 1 0 0

Supependymal giant cell astrocytoma
2 4 3 2 2 0 0 0 0 0 0 0 0 0 0 0 0 0 0 0

Superficial spreading adenocarcinoma
0 0 0 0 0 2 2 3 5 14 13 19 22 30 29 16 18 7 2 1

Superficial spreading melanoma, in situ
2 0 5 35 103 238 332 441 509 480 478 432 385 354 294 246 124 63 19 7

Superficial spreading melanoma
0 10 50 329 1064 2127 3244 3908 4329 4259 4129 39 46 3589 3116 2710 2047
     1227 596 175 74

Sweat gland adenocarcinoma
0 0 1 1 3 10 8 8 18 24 25 30 42 34 52 37 35 17 13 5

Synovial sarcoma, NOS
2 11 41 44 67 67 49 52 42 31 37 28 23 19 14 17 8 3 0 0

Synovial sarcoma, biphasic
0 6 13 21 18 14 20 13 12 11 6 9 3 5 3 1 0 1 0 0

Synovial sarcoma, epithelioid cell
0 0 0 0 0 1 1 2 0 0 1 0 2 1 0 0 0 0 0 0

Synovial sarcoma, spindle cell
1 2 9 21 14 19 21 19 13 9 15 18 4 3 4 5 1 0 0 0

Telangiectatic osteosarcoma
2 4 13 10 6 3 3 1 2 0 2 1 0 2 1 0 3 0 0 0

Teratocarcinoma
6 4 22 209 457 413 285 145 74 46 18 8 10 11 4 2 0 2 0 1

Teratoid medulloepithelioma
1 0 0 1 0 0 0 0 0 0 0 0 0 0 0 0 0 0 0 0

Teratoma with malig. transformation
0 0 0 2 1 3 2 2 4 1 5 3 4 3 3 0 1 0 0 0

Teratoma, benign
1 1 0 0 2 0 0 0 0 0 0 0 0 0 0 0 0 0 0 0

Teratoma, malignant, NOS
141 32 90 178 223 204 163 93 37 25 15 8 6 10 5 7 4 1 0 0

Let us look at a single example:

Synovial sarcoma, NOS
2 11 41 44 67 67 49 52 42 31 37 28 23 19 14 17 8 3 0 0

Corresponding ages:
0 5 10 15 20 25 30 35 40 45 50 55 60 65 70 75 80 85 90 95 (and above)

We can see that some of the distributions peak in the middle of the distributions (age 50 and above), while others seem to peak early (in children).

In Chapter 25, we will learn how to convert age distribution data into graphs, allowing us to visualize age-distribution curves.

Exercises

1. Following the instructions provided in the appendix, download the SEER data files onto your own computer.

   04/14/2009 01:50 PM 153,783,644 BREAST.TXT
   04/14/2009 01:50 PM 116,050,746 COLRECT.TXT
   04/14/2009 01:50 PM 71,956,724 DIGOTHR.TXT
   04/14/2009 01:50 PM 98,253,484 FEMGEN.TXT
   04/14/2009 01:50 PM 75,934,488 LYMYLEUK.TXT
   04/14/2009 01:50 PM 128,405,116 MALEGEN.TXT
   04/14/2009 01:50 PM 141,117,522 OTHER.TXT
   04/14/2009 01:50 PM 136,211,152 RESPIR.TXT
   04/14/2009 01:50 PM 63,487,816 URINARY.TXT

   Using Perl, Python, or Ruby, write a script that counts the aggregate number of cancer records contained in the SEER files.

2. Using Perl, Python, or Ruby, find the total number of occurrences of all types of cancer for women and for men, on separate output columns.

   Hint: You will need the byte locations for gender. The data dictionary contains this information.

3. Using Perl, Python, or Ruby, write a script that determines the list of all tumors that occur, in each and every five-year age group (0 to 95), and the frequency of occurrence of those tumors in that age group.

4. Using Perl, Python, or Ruby, find the total number of occurrences of all types of cancer for each anatomic site.

   Hint: You will need the byte locations for the anatomic site data in the SEER data records. As shown in Figure 7.1, these constitute characters 43 to 46 of the record. The codes for the anatomic sites are found in the appendix.

5. The SEER data files contain millions of tumors, but it is not clear whether the SEER data files contain examples of every tumor listed in the ICD-O. Using Perl, Python, or Ruby, write a script that determines the number of different tumors (i.e., different ICD-O codes) contained in SEER, and determine the percentage of the total number of ICD-O codes (from the ICDO nomenclature) that are accounted for in the SEER data records.

6. It is possible that when the SEER data records were written, the cancer registrars (the people who create data records from cancer cases encountered in hospitals and clinics) may have used nonstandard codes that are not included in the version of the ICD-O that we have used in these exercises (i.e., not included in the icdo3.txt file). Using Perl, Python, or Ruby, write a script that lists (if any) the nonstandard ICD-O codes in the SEER data records.

8
OMIM
The Online Mendelian Inheritance in Man

OMIM is a listing of the known inherited conditions occurring in humans. Each condition has biologic and clinical descriptions in a detailed textual narrative that includes a listing of relevant citations. The OMIM text exceeds 135 megabytes (MB) in length and can be downloaded from the National Center for Biotechnology Information's anonymous ftp site: ftp://ftp.ncbi.nih.gov and subdirectory: /repository/omim/.

OMIM is publicly available, at no cost, at

http://www.ncbi.nlm.nih.gov/sites/entrez

OMIM and Online Mendelian Inheritance in Man are registered trademarks of the Johns Hopkins University.

OMIM is an ideal and challenging corpus for testing indexing and retrieval algorithms because it contains free-text (paragraphs), structured text (lists), names (in free-text and in citations suitable for testing deidentification algorithms), gene-related terminology (names of genes, cytogenetic descriptors, proteins) and medical terms (co-morbid features of inherited diseases) and both common and obscure medical conditions. Additional information on OMIM is available at

http://www.ncbi.nlm.nih.gov/omim/

The version of OMIM available (April, 2009) is 135,191,286 bytes in length and contains 20,287 records.

The first paragraph of the first record, of over 20,000 OMIM records, is shown:

FIELD TI
100050 AARSKOG SYNDROME
FIELD TX

Grier et al. (1983) reported a father and two sons with typical Aarskog syndrome, including short stature, hypertelorism, and shawl scrotum. They tabulated the findings in 82 previous cases. An X-linked recessive inheritance has been repeatedly suggested (see 305400). The family reported by Welch (1974) had affected males in three consecutive generations. Thus, there is either genetic heterogeneity or this is an autosomal dominant with strong sex-influence and possibly ascertainment bias resulting from use of the shawl scrotum as a main criterion. Stretchable skin was present in the cases of Grier et al. (1983). Teebi et al. (1993) reported the case of an affected mother and four sons (including a pair of monozygotic twins) by two different husbands. They suggested that the manifestations were as severe in the mother as in the sons and that this suggested autosomal-dominant inheritance. Actually, the mother seemed less severely affected, which is compatible with an X-linked inheritance.

For further reading, I have described some informatics-related uses of OMIM in a previously published public domain document at:

http://www.ncbi.nlm.nih.gov/pmc/articles/PMC441395/

8.1 Collecting the OMIM Entry Terms

In the preceding section, the first paragraph of an OMIM record is shown. Note that the title of the record (the OMIM number and the list of synonymous terms for the disorder) follows the field delimiter "*FIELD* TI" and precedes the field delimiter "*FIELD* TX". The titles of records are very useful, because they allow us to link OMIM records to records in other documents and data sets that contain the OMIM number or a synonymous term. We can collect all of the OMIM title entries into a single file, that we can use later in informatics projects.

8.1.1 Script Algorithm

1. Open OMIM.
2. Parse through OMIM record by record.
3. Extract from each record the portion of text that is within the title field (*FIELD* TI).
4. Add the output information to an output file, or display it directly on the monitor.

Perl Script

```perl
#!/usr/local/bin/perl
$/ = "*RECORD*";
open (TEXT, "c\:\\big\\omim")||die"cannot";
$line = <TEXT>;
$line = " ";
$count = 0;
while ($line ne "")
    {
    $line = <TEXT>;
    $getline = $line;
    $getline =~ s/\n/ /g;
    if ($getline =~ /\*FIELD\* TI(.+)\*FIELD\* TX/m)
        {
```

```
      print lc($1) . "\n\n";
      }
   }
exit;
```

Python Script

```
#!/usr/local/bin/python
import re, string
count = 0
in_text = open('C:\\big\\omim', "r")
out_text = open("omimword.txt", "w")
clump = ""
for line in in_text:
  namematch = re.match(r'\*RECORD\*', line)
  if (namematch):
    count = count + 1
    clump = re.sub(r'\n', ' ', clump)
    fieldmatch = re.search(r'\*FIELD\* TI(.+)\*FIELD\* TX', clump)
    if fieldmatch:
      print>>out_text, string.lower(fieldmatch.group(1))
    clump = ""
  else:
    clump = clump + line
exit
```

Ruby Script

```
#!/usr/local/bin/ruby
file_in = File.open("c\:\\big\\omim")
file_in.each_line("*RECORD*") do
  |line|
  line =~ /\*FIELD\* TI(.+)\*FIELD\* TX/m
  puts $1.downcase if $1
end
exit
```

8.1.2 Analysis

The output consists of the title line for each of the approximately 20,000 OMIM records. The first four output lines are shown here:

100050 aarskog syndrome

#100070 aortic aneurysm, abdominal ;;aaa;; aaa1;; aneurysm, abdominal aortic;; abdominal aortic aneurysm arteriomegaly, included;; aneurysms, peripheral, included

100100 abdominal muscles, absence of, with urinary tract abnormality and cryptorchidism ;;prune belly syndrome;; eagle-barrett syndrome

100200 abducens palsy

This script demonstrates that if you have a text file, prepared in a consistent manner that separates entries, records, or sections, you can easily write a script that extracts, reorganizes, or transforms the specific kinds of information contained in the records.

8.2 Finding Inherited Cancer Conditions

There are over 20,000 OMIM records covering every type of inherited diseases. Occasionally, you will need to focus your attention on records that contain a particular feature of interest.

A cancer researcher may be interested in inherited syndromes that carry an increased risk of developing benign or malignant tumors (neoplasms). The OMIM records contain a special section, labeled "Oncology," listing neoplasms that may occur as part of the clinical syndrome. For example, OMIM entry 114900 contains the following text:

```
*FIELD* CS
Oncology:
Intestinal carcinoid;
Appendiceal carcinoid;
Malignant carcinoid of ileum
Inheritance:
Autosomal dominant
```

The syndrome-associated neoplasms are listed as indented terms following the "Oncology:" header. We can use the uniform record notation to extract all of the neoplasms in inherited cancer syndromes.

8.2.1 Script Algorithm

1. Open the OMIM file for reading.
2. Parse through the OMIM file record by record. Records in OMIM are delimited by the line "*RECORD*".
3. In each record, extract the text between the lines "*FIELD* TX" and "*FIELD* TI". These lines enclose the OMIM record title.
4. In each record, extract the text that follows the line, "Oncology:" and precedes a line that begins with a word that begins flush against the beginning of a new line. This happens to enclose the format of a listing of the names of neoplasms associated with an OMIM syndrome.
5. In each record, if there is an oncology record, print out the OMIM record title, followed by the list of neoplasms associated with the record.

Perl Script

```
#!/usr/local/bin/perl
$/ = "*RECORD*";
```

```perl
open (TEXT, "c\:\\big\\omim")||die"cannot";
$line = " ";
while ($line ne "")
   {
   $line = <TEXT>;
   $line =~ /\*FIELD\* TX/m;
   $front = $`;
   if ($front =~ /\*FIELD\* TI/m)
      {
      $front = $';
      }
   if ($line =~ /Oncology\:/)
      {
      $oncoterms = $';
      $oncoterms =~ /\n[A-Z]/;
      $oncoterms = $`;
      if ($oncoterms)
         {
         print lc($front) . lc($oncoterms);
         }
      }
   }
exit;
```

Python Script

```python
#!/usr/local/bin/python
import re, string
in_text = open('C:\\big\\omim', "r")
clump = ""
for line in in_text:
   namematch = re.match(r'\*RECORD\*', line)
   if namematch:
      section = re.sub(r'\n', ' ', clump)
      fieldmatch = re.search(r'\*FIELD\* TI(.+)\*FIELD\* TX', section)
      if fieldmatch:
         oncopiece = re.sub(r'\n', '@', clump)
         oncomatch = re.search(r'Oncology:(.+?)\@[A-Z]', oncopiece)
         if oncomatch:
            oncopiece = re.sub(r'@', '\n', string.lower(oncomatch.
group(1)))
            print string.lower(fieldmatch.group(1)) + " " + oncopiece +
"\n"
      clump = ""
   else:
      clump = clump + line
exit
```

Ruby Script

```
#!/usr/local/bin/ruby
file_in = File.open("c\:\\big\\omim")
file_in.each_line("*RECORD*") do
  |line|
  line =~ /\*FIELD\* TI(.+)\*FIELD\* TX/m
  entry = $1
  if entry
    line =~ /Oncology:(.+?)\n[A-Z]/m
    oncoterms = $1;
    puts entry.downcase + oncoterms.downcase if oncoterms
  end
end
exit
```

8.2.2 Analysis

The first five records of the output are shown:

102660 adamantinoma of long bones
    adamantinoma of long bones

104600 amenorrhea-galactorrhea syndrome
    pituitary adenoma

105580 anal canal carcinoma cloacogenic carcinoma, included
    anal canal squamous carcinoma

*108330 cytochrome p450, subfamily i, polypeptide 1; cyp1a1;;cytochrome p450, aromatic compound-inducible;;aryl hydrocarbon hydroxylase; ahh;;flavoprotein-linked monooxygenase;;cytochrome p1-450, dioxin-inducible;;cytochrome p1-450, inducible by 2,3,7,8-tetrachlorodibenzo-p-dioxin;;tcdd-inducible cytochrome p450;p450dx;;polycyclic aromatic compound-inducible p450
    ? high-inducibility phenotype at greater risk for bronchogenic carcinoma

%109350 gastroesophageal reflux;;ger;;gastroesophageal reflux disease; gerd1;; gastroesophageal reflux, pediatric barrett metaplasia, included;;barrett esophagus, included;;adenocarcinoma of esophagus, included
    adenocarcinoma of the esophagus risk about 10%

In just a few seconds, we can collect all of the OMIM conditions that are associated with a specific "Oncology" field that lists the neoplastic conditions associated with the disorder.

Exercises

1. Using Perl, Python, or Ruby, determine the total number of records in OMIM.

2. Some disorders are characterized by G-to-A mutations. Using Perl, Python, or Ruby, write a script that collects from OMIM all records that contain the term "G-to-A".

3. G-to-A transitions may occur as homozygous or heterozygous mutations. Using Perl, Python, or Ruby, write a script that collects, from OMIM all records that contain the term "G-to-A" and the word "heterozygous" or the word "homozygous".

4. Records that contain two closely related terms (i.e., a term that is relevant to the meaning of another term) usually occur in close proximity within the text. Using Perl, Python, or Ruby, write a script that collects from OMIM all records that contain the term "G-to-A" within 10 words (preceding or following) the word "heterozygous" or the word "homozygous".

9

PubMed

PubMed is the U.S. National Library of Medicine's public search engine for about 19 million citations from the medical literature. Each citation consists of the authors, the title, and the journal reference for each article. For the vast majority of articles, PubMed includes an abstract summarizing the research. For many articles, PubMed includes a link to the electronic version of the complete article.

You can access the PubMed data at

http://www.ncbi.nlm.nih.gov/pubmed/

or

http://www.pubmed.org/

At the same site, search engines linked to a variety of large biology databases are provided by the National Center for Biotechnology Information.

9.1 Building a Large Text Corpus of Biomedical Information

It is remarkably easy to create a large public domain text corpus for almost any medical specialty. All you need to do is to enter a PubMed query and send the results to a file on your computer's hard disk.

Here is an example of a PubMed search on the following query term:

cancer OR sarcoma OR carcinoma OR tumor OR adenocarcinoma OR neo-plasm OR lymphoma OR leukemia AND gene

This returns a list of about 360,000 citations, which can be downloaded, along with abstracts of the cited papers, and saved as a text file (Figure 9.1). Under "Format", choose "Medline" to produce an output that provides detailed information on each record. The "Choose Destination" box in the upper-right corner of the image permits the user to download the complete search results. By selecting "File", the results will be downloaded into a file on your hard drive.

The downloaded file of the returned citation list exceeds 200 megabytes (MB) in length.

The U.S. Copyright office stipulates that the names of authors and the titles of works are excluded from copyright. This makes sense because if you cannot freely publish the names of authors or the titles of their works, how would anyone know that

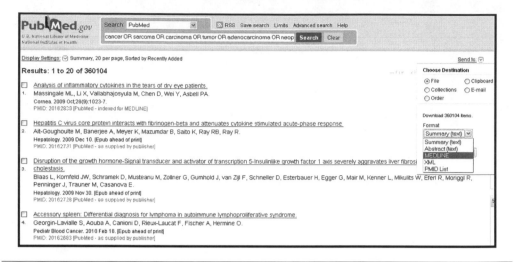

Figure 9.1 PubMed screen of query results, showing the first page of returned citations.

the work exists? The abstracts, included in PubMed downloads, are covered by copyright, and cannot be republished. If you want to have a corpus of text extracted from a PubMed download that you can freely distribute, you will want to extract the titles.

9.1.1 Script Algorithm

1. Open the PubMed download file.
2. Open an output file.
3. PubMed download files contain records that consistently begin with "PMID- ". Parse through the PubMed download file, record by record, using "PMID- " as the record separator.
4. Within the record, the title field is preceded by "TI - ", and the title ends with a newline character followed by another field designator, such as the abstract field designator, "AB - ".

      For example:

          TI—A Wnt Survival Guide: From Flies to Human Disease.

          AB—It has been two decades since investigators discovered the...

      From each record, extract the text that lies between the title field designator and the next field designator.

5. Convert the title to lowercase.
6. Clean the title line by removing nonalphanumeric characters, extra spaces, possessive markers ("'s"), and the plural forms of tumor names.
7. Write titles to an external output file.

Perl Script

```perl
#!/usr/local/bin/perl
$/ = "PMID- ";
```

```
open(TEXT,"c\:\\big\\cancer_gene_pubmed.txt")||die"cannot";
open(OUT,">cancer_gene_titles.txt");
$line = " ";
while ($line ne "")
  {
  $line = <TEXT>;
  if ($line =~ /\nTI  \-(.+)\n[A-Z]{2}  \-/)
    {
    $title = $1;
    $title = lc($title);
    $title =~ s/\'s//g;
    $title =~ s/\W/ /g;
    $title =~ s/omas/oma/g;
    $title =~ s/tumour/tumor/g;
    $title =~ s/\n/ /g;
    $title =~ s/^ +//;
    $title =~ s/ +$//;
    $title =~ s/ +/ /g;
    next if ($title !~ /[a-z]+/);
    print OUT "$title\n";
    }
  }
exit;
```

Python Script

```python
#!/usr/local/bin/python
import string, re
in_text = open("c:\\big\\cancer_gene_pubmed.txt", "r")
out_text = open("cancer_gene_titles.txt", "w")
clump = ""
for line in in_text:
  record_match = re.match(r'PMID- ', line)
  if record_match:
    title_match = re.search(r'\nTI  -(.+)\n[A-Z]{2}  -', clump)
    if title_match:
      title = title_match.group(1)
      title = string.lower(title)
      title = re.sub(r'\'s', "", title)
      title = re.sub(r'\W', " ", title)
      title = re.sub(r'omas', "oma", title)
      title = re.sub(r'tumour', "tumor", title)
      title = re.sub(r'\n', " ", title)
      title = string.rstrip(title)
      title = string.lstrip(title)
      title = re.sub(r' +', " ", title)
      text_match = re.search(r'[a-z]+', title)
      if not text_match:
        clump = ""
```

```
        continue
      print>>out_text, title
    clump = ""
  else:
    clump = clump + line
exit
```

Ruby Script

```
#!/usr/local/bin/ruby
$/ = "PMID- "
in_text = File.open("c:/big/cancer_gene_pubmed.txt", "r")
out_text = File.open("cancer_gene_titles.txt", "w")
in_text.each_line do
  |line|
  if (line =~ /\nTI  \-(.+)\n[A-Z]{2}  \-/)
    title = $1.downcase
    title.gsub!(/\'s/, "") if title =~ /\'s/
    title.gsub!(/\W/, " ") if title =~ /\W/
    title.gsub!(/omas/, "oma") if title =~ /omas/
    title.gsub!(/tumour/, "tumor") if title =~ /tumour/
    title.gsub!(/\n/, " ") if title =~ /\n/
    title = title.strip
    title.gsub!(/ +/, " ") if title =~ / +/
    next if (title !~ /[a-z]+/)
    out_text.puts title
  end
end
exit
```

9.1.2 Analysis

The output is a public domain file consisting of lowercase reference titles, without punctuation (Figure 9.2).

A public domain file of titles related to cancer genes and tumors is available at

http://www.julesberman.info/book/cancer_gene_titles.txt

We will use this file in the next section.

9.2 Creating a List of Doublets from a PubMed Corpus

Autocoding is a specialized form of machine translation. The general idea behind machine translation is that computers have the patience, stamina, and speed to quickly parse through gigabytes of text, matching text terms with equivalent terms from an external vocabulary. Human translators often scoff at the output of machine translators, noting the high rate of comical errors. An often-cited, perhaps apocryphal,

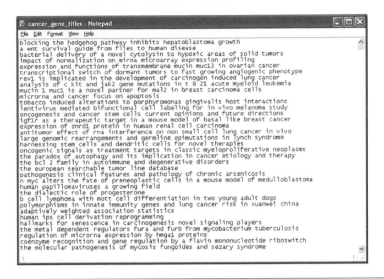

Figure 9.2   Some of the lowercase, unpunctuated PubMed article titles in the output file.

example of poor machine translation is the English to Russian transformation of "out of sight, out of mind" to the Russian equivalent of "invisible idiot."

Despite limitations, machine translation is the only way to transform gigabytes and terabytes of text. As long as clinicians, pathologists, radiologists, nurses, and scientists continue to type messages, reports, manuscripts, and notes into electronic documents, we will need computers to parse and organize the resulting text.

One of the many problems in the field of machine translation is that expressions (multiword terms) convey ideas that transcend the meanings of the individual words in the expression. Consider the following sentence:

"The ciliary body produces aqueous humor."

The example sentence has unambiguous meaning to anatomists, but each word in the sentence can have many different meanings. "Ciliary" is a common medical word, and usually refers to the action of cilia. Cilia are found throughout the respiratory and GI tract and have an important role locomoting particulate matter. The word "body" almost always refers to the human body. The term "ciliary body" should (but does not) refer to the action of cilia that move human bodies from place to place. The word "aqueous" always refers to water. Humor relates to something being funny. The term "aqueous humor" should (but does not) relate to something that is funny by virtue of its use of water (as in squirting someone in the face with a trick flower). Actually, "ciliary body" and "aqueous humor" are each examples of medical doublets whose meanings are specific and contextually constant (i.e., always mean one thing). Furthermore, the meanings of the doublets cannot be reliably determined from the individual words that constitute the doublet, because the individual words have several different meanings. Basically, you either know the correct meaning of the doublet or you don't.

Any sentence can be examined by parsing it into an array of intercalated doublets:

"The ciliary, ciliary body, body produces, produces aqueous, aqueous humor."

The important concepts in the sentence are contained in two doublets (ciliary body and aqueous humor). A nomenclature containing these doublets would allow us to extract and index these two medical concepts. A nomenclature consisting of single words might miss the contextual meaning of the doublets.

What if the term were larger than a doublet? Consider the tumor "orbital alveolar rhabdomyosarcoma." The individual words can be misleading. This orbital tumor is not from outer space, and the alveolar tumor is not from the lung. The three-word term describes a sarcoma arising from the orbit of the eye that has a morphology character-ized by tiny spaces of a size and shape as may occur in glands (alveoli). The term "orbital alveolar rhabdomyosarcoma" can be parsed as "orbital alveolar, alveolar rhabdomyosar-coma." Why is this any better than parsing the term into individual words, as in "orbital, alveolar, rhabdomyosarcoma"? The doublets, unlike the single words, are highly specific terms that are unlikely to occur in association with more than a few specific concepts.

What if a term is a singleton (a single word term)? Very few medical terms are singletons. In "The developmental lineage classification and taxonomy of neoplasms" there are about 130,000 unique terms for neoplasms. All but a few hundred of these are multiword terms.

The text in this section is an excerpt from a public domain document (Berman J. J., Doublet method for very fast autocoding. *BMC Med Inform Decis Mak*, 4:16, 2004).

We will be using doublets in later chapters, for a variety of different informatics projects. For all these projects, we will need to create an electronic list of the doublets contained in a text corpus. Let us create a doublet list from the PubMed corpus pre-pared in the previous section.

9.2.1 Script Algorithm

1. Open a text file. In this case, we will use cancer_gene_titles.txt, a list of 28,102 titles prepared in Section 9.1. Because the titles of copyrighted works are exempted from copyright restrictions, the file belongs to the public domain. A copy of the file can be downloaded at

    http://www.julesberman.info/book/cancer_gene_titles.txt

2. Parse through the file, line by line.
3. For each line of the file, parse through every doublet on the line. This means, looking at each two-word doublet consisting of each word in the line, with the word that follows.
4. As each doublet is encountered, add the doublet to a dictionary object. The dic-tionary object will have doublets as keys and the empty string, "", as the value

for each doublet. Some doublets will occur more than once in the text. A replicate doublet will generate a preexisting key–value pair and will not increase the size of the dictionary object.

5. After the text is parsed, print out the keys of the dictionary object to an external file.

Perl Script

```perl
#!/usr/local/bin/perl
open (TEXT, "c\:\\big\\cancer_gene_titles.txt")||die"Can't open file";
open (OUT, ">doubs.txt")||die"Can't open file";
$line = " ";
while ($line ne "")
   {
   $line = <TEXT>;
   $line =~ s/\n//;
   @hoparray = split(/ /,$line);
   for ($i=0;$i<(scalar(@hoparray)-1);$i++)
      {
      $doublet = "$hoparray[$i] $hoparray[$i+1]";
      if ($doublet =~ /^[a-z]+ [a-z]+$/)
         {
         $doubhash{$doublet}="";
         }
      }
   }
while ((my $key, my $value) = each(%doubhash))
   {
   print OUT "$key\n";
   }
exit;
```

Python Script

```python
#!/usr/local/bin/python
import re
import string
intext = open("c:\\big\\cancer_gene_titles.txt", "r")
outtext = open("doubs.txt", "w")
doubhash = {}
doublet = ""
doub_match = re.compile(r'^[a-z]+ [a-z]+$')
for line in intext:
   line = line.strip()
   line_array = re.split(r'\s+',line)
   line_array.append("")
   for i in range(len(line_array)-1):
     doublet = line_array[i] + " " + line_array[i+1]
     if doub_match.search(doublet):
       doubhash[doublet]=""
```

```
for k,v in doubhash.iteritems():
    print>>outtext, k
exit
```

Ruby Script

```
#!/usr/local/bin/ruby
intext = File.open("c:/big/cancer_gene_titles.txt", "r")
outtext = File.open("doubs.txt", "w")
doubhash = Hash.new(0)
line_array = Array.new(0)
while record = intext.gets
  oldword = ""
  line_array = record.chomp.strip.split(/\s+/)
  line_array.each do
    |word|
    doublet = [oldword, word].join(" ")
    oldword = word
    next unless (doublet =~ /^[a-z]+\s[a-z]+$/)
    doubhash[doublet] = ""
    end
end
doubhash.each {|k,v| outtext.puts k }
exit
```

9.2.2 Analysis

The output file, doubs.txt is 1,266,865 bytes in length and contains 77,257 doublets. The file is available for download at

http://www.julesberman.info/book/doubs.txt

A few doublet entries from the output file are shown:

development of
favorable neuroblastoma
show evidence
carcinoma atypical
mediastinum a
localized hepatic
combining microarray
neoplastic metastasis
pathophysiology of
erbb receptor
illuminate intersection
by knock

ct antigen

candidate pro

hemangioma after

proper activation

lipoproteins and

of granular

the microscope

When the original text has no identifying, misspelled, profane, or otherwise objectionable text, the resulting doublets can be used as "safe" for inclusion in confidential text (see Chapter 15). In this case, we extracted doublets from a corpus consisting of the titles of scientific articles. These titles would not be expected to contain identifying or objectionable doublets.

9.3 Downloading Gene Synonyms from PubMed

At the PubMed site, select "Gene" as your Search Engine, and enter "geneid" as your query. PubMed will return a large set of geneid entries (230,201 in the example), which you can download (Figure 9.3).

The records serve as a text corpus from which you can extract a gene nomenclature.

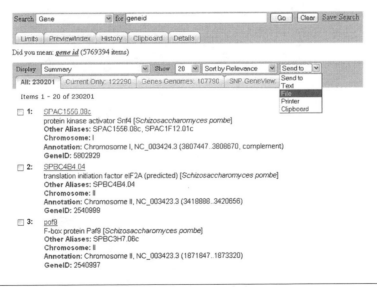

Figure 9.3 A large data set of gene names and related information can be obtained by searching in the "Gene" database, and entering "geneid") as your query term. Three entries are shown. Click on the "Send to" box and select "File." The download produces a file exceeding 42 MB.

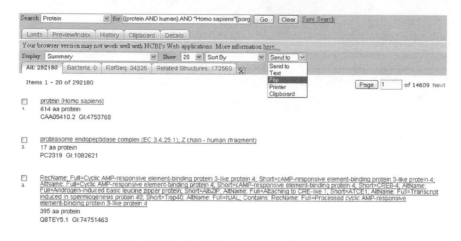

Figure 9.4   The first three entries of a PubMed search. All 292,180 query results can be downloaded by clicking on the "Send to" list box, in the upper right corner, and choosing "File."

9.4 Downloading Protein Synonyms from PubMed

Select the "Protein" database, and enter the query (Figure 9.4):

> ((protein AND human) AND "Homo sapiens"[porgn:__txid9606]) AND
> "Homo sapiens"[porgn:__txid9606]

In this example, the results yielded 292,180 entries. It is easy to see that the output file can be easily parsed, and protein information can be integrated with any other data sets that contain information on any virtually any protein.

Proteins, along with their ontologic relationships, are also available for download from GO, the Gene Ontology project.

> http://www.geneontology.org/ontology/gene_ontology_edit.obo

This file is curated and updates are frequent. Currently (2009), the file exceeds 16 MB. The entries in the GO databases are formatted as shown in the following example:

> [Term]
> id: GO:0000001
> name: mitochondrion inheritance
> namespace: biological_process
> def: "The distribution of mitochondria, including the mitochondrial genome, into daughter cells after mitosis or meiosis, mediated by interactions between mitochondria and the cytoskeleton." [GOC:mcc, PMID:10873824, PMID:11389764]
> synonym: "mitochondrial inheritance" EXACT []
> is_a: GO:0048308 ! organelle inheritance
> is_a: GO:0048311 ! mitochondrion distribution

Exercises

1. Our script that extracts doublets from a large PubMed corpus is imperfect. Because it parses line by line, instead of sentence by sentence or paragraph by paragraph, it cannot find doublets that flank the end of one line and the beginning of the next line. Using your preferred language, write a script that overcomes this deficit.

2. Write a script that parses through a paragraph of text, extracting every sub-sequence of words, of every possible number (i.e., the individual words in the paragraph plus the two-word phrases in the paragraph plus the three-word phrases and so on until you have the sequence of words that comprise the entire paragraph.

3. There are more records in the protein corpus (286,280) than in the gene corpus (214,763). The protein corpus only includes human proteins, whereas the gene corpus includes genes from many different organisms. Using Perl, Python, or Ruby, write a script that parses through the geneid data file, collecting only those genes that occur in humans. What is the ratio of the number of different human genes compared with number of different human proteins?

4. Modify the script from Section 9.2.2, to create a list of doublets, with the number of occurrences of each doublet in the corpus preceding the doublet term. Order the doublets in an output file by descending frequency. Inspect the output file. Are there any generalizations you can make regarding the potential utility of terms near the top of the file (the most frequently occurring terms), the bottom of the file (terms that occur just once or twice in the corpus), and the middle of the file (terms that occur occasionally throughout the corpus)?

10

TAXONOMY

Taxonomy.dat is a large, publicly available list of organisms. The file is available from the European Bioinformatics Institute (EBI). It contains over 580,000 species:

A sample record in Taxonomy.dat
ID : 50
PARENT ID : 49
RANK: genus
GC ID : 11
SCIENTIFIC NAME : Chondromyces
SYNONYM : Polycephalum
SYNONYM : Myxobotrys
SYNONYM : Chondromyces Berkeley and Curtis 1874
SYNONYM : "Polycephalum" Kalchbrenner and Cooke 1880
SYNONYM : "Myxobotrys" Zukal 1896
MISSPELLING : Chrondromyces

The Taxonomy.dat file exceeds 100 megabytes (MB) in length.

The Taxonomy.dat file is available for public download through anonymous ftp.

ftp://ftp.ebi.ac.uk/pub/databases/taxonomy/

More information on the Taxonomy.dat file is found at

http://www.ebi.ac.uk/msd-srv/docs/dbdoc/ref_taxonomy.html

Notice that the sample entry (above) provides an ID number for the entry organism, Chondromyces, and for its parent class. Since every organism and class has a parent, you can write a script that reconstructs the full phylogenetic lineage for any entry in Taxonomy.dat.

10.1 Finding a Taxonomic Hierarchy

The script parses through Taxonomy.dat, build a hash of all the child–parent relationships, then re-parse the file, building the phylogenetic lineage of each organism using the child-parent hash that was built in the first pass.

10.1.1 Script Algorithm

1. Open the Taxonomy.dat file for reading.
2. Records in the Taxonomy.dat file are separated by a "//" occurring at the beginning of a line. Parse through the Taxonomy.dat file, line by line.
3. Each record of taxonomy contains the name of the organism featured in the record, the taxonomy code for the organism, and the taxonomy code for the parent class of the organism. As each record is parsed, pass key–value pairs into two dictionary objects. One dictionary object holds the code number of the record's organism as its key and the code number of the organism's parent class as its value. The other dictionary object holds the code number of the record's organism as its key and the scientific name of the organism as its hash.
4. After the Taxonomy.dat file has been parsed, the two dictionary object contain all the information needed to reconstruct the full taxonomic hierarchy for any organism.
5. Close and open the Taxonomy.dat file, so that every record can be parsed once more. We will use the second parse to collect the names of organisms in the taxonomy.
6. As the Taxonomy.dat file is parsed, extract the name of the organism of record, determine that record's scientific name, and print it out. Use the dictionary containing the parent code to determine the direct parent of the organism. Use the dictionary object containing the scientific names of organisms to determine the scientific name of the parent organism. Print the parent code and the parent scientific name. Repeat this until the hierarchy is exhausted (i.e., when the lineage reaches "root.")
7. Repeat this for each record in Taxonomy.dat.

Perl Script

```
#!/usr/local/bin/perl
open(TAXO, "taxonomy.dat");
open(OUT, ">taxo.txt");
$/ = "//";
$line = " ";
while ($line ne "")
  {
  $line = <TAXO>;
  $line =~ /\nID +\: *([0-9]+) *\n/;
  $id_name = $1;
  $line =~ /\nPARENT ID +\: *([0-9]+) *\n/;
  $parent_id_name = $1;
  $parenthash{$id_name} = $parent_id_name;
  $line =~ /\nSCIENTIFIC NAME +\: *([^\n]+) *\n/;
  $scientific_name = $1;
  $namehash{$id_name} = $scientific_name;
  }
```

```
close(TAXO);
open(TAXO, "taxonomy.dat");
$line = " ";
while ($line ne "")
   {
   $line = <TAXO>;
   $getline = $line;
   $getline =~ s/\/\///o;
   print OUT $getline . "HIERARCHY\n";
   $line =~ /\nID +\: *([0-9]+) *\n/;
   $id_name = $1;
   for(1..30)
      {
      print OUT "$namehash{$id_name}\n";
      $id_name = $parenthash{$id_name};
      last if ($namehash{$id_name} eq "root");
      }
   print OUT "//";
   }
exit;
```

Python Script

```python
#!/usr/local/bin/python
import re
intext = open("taxonomy.dat", "r")
outtext = open("taxo.txt", "w")
parenthash = {}
namehash = {}
cum_line = ""
childnumber = ""
parentnumber = ""
child_match = re.compile('ID\s+\:\s*(\d+)\s*')
parent_match = re.compile('PARENT ID\s+\:\s*(\d+)\s*')
name_match = re.compile('SCIENTIFIC NAME\s+\:\s*([^\n]+)\s*')
end_match = re.compile('\/\/')
for line in intext:
   p = end_match.search(line)
   if p:
      m = child_match.search(cum_line)
      if m:
         childnumber = m.group(1)
      x = parent_match.search(cum_line)
      if x:
         parentnumber = x.group(1)
      parenthash[childnumber] = parentnumber
      y = name_match.search(cum_line)
      if y:
         scientific_name = y.group(1)
      namehash[childnumber] = scientific_name
```

```
      cum_line = ""
      continue
    else:
      cum_line = cum_line + line
cum_line = ""
intext.close
intext = open("taxonomy.dat", "r")
for line in intext:
  p = end_match.search(line)
  if p:
    print>>outtext, cum_line + "HIERARCHY"
    z = child_match.search(cum_line)
    if z:
      id_name = z.group(1)
      for i in range(30):
        if namehash.has_key(id_name):
          print>>outtext, namehash[id_name]
        if parenthash.has_key(id_name):
          id_name = parenthash[id_name]
    print>>outtext, "//"
    cum_line = ""
    continue
  else:
    cum_line = cum_line + line
cum_line = ""
exit
```

Ruby Script

```ruby
#!/usr/local/bin/ruby
intext = File.open("taxonomy.dat", "r")
outtext = File.open("taxo.txt", "w")
parenthash = Hash.new()
namehash = Hash.new()
intext.each_line("//") do
  |line|
  line =~ /\nID\s+\:\s*([0-9]+)\s*\n/
  child_id = $1
  line =~ /\nPARENT ID\s+\:\s*([0-9]+)\s*\n/
  parent_id = $1
  parenthash[child_id] = parent_id
  line =~ /\nSCIENTIFIC NAME\s+\:\s*([^\n]+)\s*\n/
  scientific_name = $1
  namehash[child_id] = scientific_name
end
intext.close
intext = File.open("taxonomy.dat", "r")
intext.each_line("//") do
  |line|
  getline = line
```

```
  getline.sub!(/\/\//,"")
  outtext.puts(getline, "HIERARCHY")
  line =~ /\nID\s+\:\s*([0-9]+)\s*\n/
  id_name = $1
  (1..30).each do
    outtext.puts(namehash[id_name])
    id_name = parenthash[id_name]
    break if namehash[id_name].nil?
  end
  outtext.print("//")
end
exit
```

10.1.2 Analysis

These scripts produce an output file, taxo.txt, that exceeds 224 MB in length. The output consists of the taxonomic entries from Taxonomy.dat, along with the phylogentic lineage for each organism.

It takes under a minute to execute these scripts on a desktop computer running at 2.5 GHz with 512 MB RAM.

Sample output for the phylogenetic hierarchy for *Homo sapiens*:

9606 Homo sapiens
9605 Homo
207598 Homo/Pan/Gorilla group
9604 Hominidae
314295 Hominoidea
9526 Catarrhini
314293 Simiiformes
376913 Haplorrhini
9443 Primates
314146 Euarchontoglires
9347 Eutheria
32525 Theria
40674 Mammalia
32524 Amniota
32523 Tetrapoda
8287 Sarcopterygii
117571 Euteleostomi
117570 Teleostomi
7776 Gnathostomata
7742 Vertebrata
89593 Craniata
7711 Chordata

33511 Deuterostomia
33316 Coelomata
33213 Bilateria
6072 Eumetazoa
33208 Metazoa
33154 Fungi/Metazoa group
2759 Eukaryota
131567 cellular organisms
1 root

A Web site that automatically generates the phylogenetic lineage of any entered species (listed in Taxonomy.dat) is available:

http://www.julesberman.info/post.htm

10.2 Finding the Restricted Classes of Human Infectious Pathogens

There are just a few hundred infectious organisms that produce diseases in humans. A list of the generally accepted human pathogens is available for download at

http://www.julesberman.info/book/infect.txt

For every infectious pathogen of man, we can determine its taxonomic lineage. By grouping pathogens by their lineage, we reach several important results.

1. We can quantify which phyla are the most dangerous to man (and these would be the small number of phyla known to contain pathogenic organisms).
2. We can simplify the task of learning the biology of individual species (because we can generalize a great number of species under the inherited properities of a small number of biological classes).
3. We can better guess the potential pathogenicity of organisms that are, for the first time, isolated from human lesions (because organisms of the same class are apt to cause disease through similar biological mechanisms).
4. We can better guess the drugs that will be effective against a new human pathogen (if we know the class of organisms in which it belongs).

10.2.1 Script Algorithm

1. Open Taxonomy.dat and parse through every organism record (there are 583,049 records in the version of taxonomy that I downloaded on June 5, 2009).
2. As each record is parsed, contribute key–value pairs to four different dictionary objects:
   a. The dictionary object whose keys are the standard identifiers for the organisms or class names. Identifiers are numeric. The values will be the

scientific names that correspond to the numeric identifiers. The key–value pair for us is 9606/Homo sapiens. Humans are identified by "9606" in a wide variety of biomedical databases.

b. The reverse dictionary object whose keys are the scientific name of the organisms or class names and whose values are the corresponding numeric identifiers.

c. A parent dictionary object whose keys are the identifiers of the organisms or classes and whose values are the names of the parent of the organism or class (provided in each Taxonomy.dat record).

d. The dictionary object whose keys are the numeric id of the organism or class and whose values are the corresponding ranks of the organism or class. Every record in Taxonomy.dat corresponds to the name of an organism (the species name) or to the name of a class of organisms. The classes are given scientific ranks (such as domain, kingdom, phylum, class, order, family, genus, species).

3. When the Taxonomy.dat file has been entirely parsed, and all four dictionary objects have been built, we can reconstruct the entire lineage of any organism by iterating over the name of the parent class for the entry, until the parental lineage is exhausted. If we are interested in any particular ranking (the organism's phyla in this case), we can stop the iteration loop when an ancestor with rank "phyla" is encountered.

4. Open the infect.txt file, containing the list of most infectious organisms of humans, with one organism assigned to each line of the file.

5. Parse through the infect.txt file, looping through the lineage of each organism by determining successive parent names.

6. When a parent name is encountered whose rank is "phyla," increment the value of a dictionary object whose keys are the names of the phyla and whose values are the total number of times the phyla name has been encountered in the lineages of the list of infectious organisms.

7. Print the lineages of each infectious organism to an output file.

8. Print the phyla dictionary object, with the value (number of occurrences of the phyla) followed by the name of the phyla, listed from most frequently occurring phyla to least frequently occurring phyla.

Perl Script

```
#!/usr/local/bin/perl
open(TAXO, "c\:\\ftp\\taxonomy.dat");
$/ = "//";
$line = " ";
while ($line ne "")
  {
  $line = <TAXO>;
```

```perl
   $line =~ /\nID +\: *([0-9]+) *\n/;
   $id_name = $1;
   $line =~ /\nPARENT ID +\: *([0-9]+) *\n/;
   $parent_id_name = $1;
   $parenthash{$id_name} = $parent_id_name;
   $line =~ /\nSCIENTIFIC NAME +\: *([^\n]+) *\n/;
   $scientific_name = lc($1);
   $namehash{$id_name} = $scientific_name;
   $termhash{$scientific_name} = $id_name;
   $line =~ /\nRANK *\: *([a-z]+) *\n/;
   $rank = $1;
   $rankhash{$id_name} = $rank;
   }
close(TAXO);
open(TAXO, "c\:\\ftp\\infect.txt");
open(OUT, ">phylum.txt");
$/ = "\n";
$line = " ";
while ($line ne "")
   {
   $line = <TAXO>;
   $scientific_name = lc($line);
   $scientific_name =~ s/\n$//o;
   my @darwin;
   next unless (exists($termhash{$scientific_name}));
   $id_name = $termhash{$scientific_name};
   for(1..30)
      {
      push(@darwin, ucfirst($namehash{$id_name}));
      if ($rankhash{$id_name} eq "phylum")
         {
         $term_name = $namehash{$id_name};
         $phyla_count{$term_name}++;
         last;
         }
      $id_name = $parenthash{$id_name};
      }
   @darwin = reverse(@darwin);
   print OUT join(":", @darwin);
   print OUT "\n";
   }
while ((my $key, my $value) = each(%phyla_count))
   {
   if ($value > 0)
      {
      $value = "000" . $value;
      $value = substr($value, -2, 2);
      push(@count_array, "$value $key");
      }
   }
```

```
@count_array = reverse(sort(@count_array));
print join("\n",@count_array);
exit;
```

Python Script

```python
#!/usr/local/bin/python
import re, string
in_file = open("c:/ftp/taxonomy.dat", "r")
parenthash = {}
namehash = {}
termhash = {}
rankhash = {}
phyla_count = {}
old_line = "\n"
for line in in_file:
  line_end = re.match(r'//',line)
  if line_end:
    line = old_line
    id_match = re.search(r'\nID +\: *([0-9]+) *\n', line)
    if id_match:
      id_name = id_match.group(1)
    parent_match = re.search(r'\nPARENT ID +\: *([0-9]+) *\n', line)
    if parent_match:
      parent_id_name = parent_match.group(1)
      parenthash[id_name] = parent_id_name
    science_match = re.search(r'\nSCIENTIFIC NAME +\: *([^\n]+)
*\n', line)
    if science_match:
      scientific_name = science_match.group(1)
      scientific_name = string.lower(scientific_name)
      namehash[id_name] = scientific_name
      termhash[scientific_name] = id_name
    rank_match = re.search(r'\nRANK *\: *([a-z ]+[a-z]) *\n', line)
    if rank_match:
      rank = rank_match.group(1)
      rankhash[id_name] = rank
    old_line = "\n"
  else:
    old_line = old_line + line
in_file.close()
in_file = open("c:/ftp/infect.txt", "r")
out_file = open("phylum.txt", "w")
for line in in_file:
  scientific_name = string.lower(line)
  scientific_name = string.rstrip(scientific_name)
  darwin = []
  if not (termhash.has_key(scientific_name)):
    continue
  id_name = termhash[scientific_name]
```

```python
    for iterations in range(30):
      if namehash.has_key(id_name):
        darwin.append(namehash[id_name])
      if rankhash.has_key(id_name):
        if (rankhash[id_name] == "phylum"):
          term_name = namehash[id_name]
          if phyla_count.has_key(term_name):
            phyla_count[term_name] = int(phyla_count[term_name]) + 1
          else:
            phyla_count[term_name] = 1
          break
      id_name = parenthash[id_name]
  darwin.sort()
  darwin.reverse()
  print>>out_file, ':'.join(darwin)
count_array = []
for key,value in phyla_count.iteritems():
  if int(value) > 0:
    value = "000" + str(value)
    value = value[-2:]
    count_array.append(value + " " + key)
count_array.sort()
count_array.reverse()
print '\n'.join(count_array)
exit
```

Ruby Script

```ruby
#!/usr/local/bin/ruby
in_file = File.open("c:/ftp/taxonomy.dat", "r")
$/ = "//"
parenthash = {}
namehash = {}
termhash = {}
rankhash = {}
phyla_count = {}
in_file.each_line do
  |line|
  line =~ /\nID +\: *([0-9]+) *\n/;
  id_name = $1;
  line =~ /\nPARENT ID +\: *([0-9]+) *\n/
  parent_id_name = $1
  parenthash[id_name] = parent_id_name
  line =~ /\nSCIENTIFIC NAME +\: *([^\n]+) *\n/
  if $1
    scientific_name = $1.downcase
    namehash[id_name] = scientific_name
    termhash[scientific_name] = id_name
  end
  line =~ /\nRANK *\: *([a-z]+) *\n/
```

```
    if $1
      rank = $1;
      rankhash[id_name] = rank;
    end
end
in_file.close
in_file = File.open("c:/ftp/infect.txt", "r")
out_file = File.open("phylum.txt", "w")
$/ = "\n"
in_file.each_line do
  |line|
  scientific_name = line.downcase
  scientific_name.chomp!
  darwin = []
  next unless termhash.has_key? scientific_name
  id_name = termhash[scientific_name]
  (1..30).each do
    |n|
    darwin.push(namehash[id_name])
    if rankhash[id_name] == "phylum"
       term_name = namehash[id_name]
       phyla_count[term_name] = phyla_count[term_name].to_i + 1
       break
    end
    id_name - parenthash[id_name]
  end
  out_file.puts darwin.reverse.join(":")
end
count_array = []
phyla_count.each do
  |key,value|
  if value > 0
    value = "000" + value.to_s
    value = value.slice(-2, 2)
    count_array.push(value + " " + key)
  end
end
puts count_array.sort.reverse.join("\n")
exit
```

10.2.2 Analysis

Here is the output consisting of the phyla accounting for infectious diseases in humans, and their frequency of occurrence in the list of pathogens.

63 proteobacteria
35 actinobacteria
27 nematoda

22 platyhelminthes
19 ascomycota
13 firmicutes
08 apicomplexa
05 spirochaetes
04 tenericutes
02 microsporidia
02 fusobacteria
02 chlamydiae
02 bacteroidetes
01 chlorophyta

The script also produces a file, with the lineage of each organism included in the infect.txt file, truncated at the phylum rank (phylum.txt)

A sample line from the output file, phylum.txt, is shown here:

Proteobacteria:Gammaproteobacteria:Legionellales:Legionellaceae:Legionella:
    Legionella pneumophila

Not all of the lines in the file have lineages that stop at the "phylum" level. Here is another line from phylum.txt:

:Root::Root::Root::Root::Root::Root::Root::Root::Root::Root::Cellular organis
    ms:Eukaryota:Euglenozoa:Kinetoplastida:Trypanosomatidae:Leishmania:Le
    ishmania:Leishmania aethiopica species complex:Leishmania aethiopica

What happened here? The script loops through the parental lineage up to 30 times, or until an ancestor with the rank of "phyla" is encountered. Most lineages in Taxonomy.dat have an ancestor with the rank of "phyla," but when a lineage excludes any ancestor with a "phyla" rank, the loop continues 30 times, producing an unattractive output.

How could we have prevented this problem? The only remedy is to go into the Taxonomy.dat file and try to determine why the organism lacked an ancestor of rank "phyla". This may require expertise in the field of taxonomy. It may require the investigator to contact the curators of the Taxonomy.dat file. It may require a qualifying remark indicating that the phyla counts are somewhat inaccurate. The moral of the story is that medical informatics often requires biological expertise in addition to programming skills.

Exercises

1. Write a script that computes the closest common ancestor for any two selected organisms (in the taxonomy).
2. Apicomplexa is a class of Protoctists that can burrow into other organisms and live as symbiotrophs. Not surprisingly, class Apicomplexa contains some of the most dangerous pathogens of humans and other animals. Using Perl,

Python, or Ruby, write a script that collects all of the organisms that are descendants of Class Apicomplex, and listed in Taxonomy.dat.

3. The pathal.txt file contains the names of 396 organisms that cause diseases in humans. The file is available at

http://www.julesberman.info/book/pathal.txt

Here are the first few lines of the file
   Acanthamoeba castellanii
   Actinobacillus actinomycetemcomitans
   Actinomadura madurae
   Actinomadura pelletieri
   Actinomyces gerencseriae
   Actinomyces israelii
   Actinomyces pyogenes
   Alcaligenes species
   Ancylostoma duodenale
   Angiostrongylus cantonensis
   Angiostrongylus costaricensis
   Anisakis simplex
   Arcanobacterium haemolyticum, Corynebacterium haemolyticum
   Arenaviridae, LCM-Lassa-virus complex Ippy
   Arenaviridae, LCM-Lassa-virus complex Lassa fever
   Arenaviridae, LCM-Lassa-virus complex Lujo
   Arenaviridae, LCM-Lassa-virus complex Lymphocytic choriomeningitis
The first word in each line is the genus of the species. Using Perl, Python, or Ruby, write a script that builds the taxonomic hierarchy for each genus in the pathal.txt file (i.e., the hierarchy for Acanthamoeba, Actinobacillus, Actinomadura, Actinomyces, Alcaligenes, etc.), that is found in the Taxonomy.dat file. This project will yield all of the major and minor taxonomic classes for the pathogens of humans.

4. Among the angiosperms (flowering plants), the most important class division are the monocots (Liliopsida) and the dicots (eudicotyledons). Using Perl, Python, or Ruby, write a script that lists all of the plants that are monocots and, in another list, all the plants that are dicots.

5. Using Perl, Python, or Ruby, write a script that produces the lineage for each organism in Taxonomy.dat, with each line consisting of the lineage list.
   Hint: There are 583,049 organisms in Taxonomy.dat, and this will be the number of lines expected in the output file.
An example of the one line from the output file is:
   cellular organisms:Eukaryota:Alveolata:Apicomplexa:Aconoidasida
   :Haemosporida:Plasmodium:Plasmodium (Laverania):Plasmodium
   falciparum:Plasmodium falciparum FCM17/Senegal
The output will look something similar to Figure 10.1.

6. Modify your script from Exercise 5 to include the rank of each listed ancestor for each species.

```
:root
:cellular organisms:Bacteria
:cellular organisms:Bacteria:Proteobacteria:Alphaproteobacteria:Rhizobiales:Xanthobacteraceae:Azorhizobi
:cellular organisms:Bacteria:Proteobacteria:Alphaproteobacteria:Rhizobiales:Xanthobacteraceae:Azorhizobi
:cellular organisms:Bacteria:Proteobacteria:Gammaproteobacteria:Enterobacteriales:Enterobacteriaceae:Buc
:cellular organisms:Bacteria:Proteobacteria:Gammaproteobacteria:Pseudomonadales:Pseudomonadaceae:Cellvib
:cellular organisms:Bacteria:Proteobacteria:Gammaproteobacteria:Pseudomonadales:Pseudomonadaceae:Cellvib
:cellular organisms:Bacteria:Dictyoglomi:Dictyoglomia:Dictyoglomales:Dictyoglomaceae:Dictyoglomus
:cellular organisms:Bacteria:Dictyoglomi:Dictyoglomia:Dictyoglomales:Dictyoglomaceae:Dictyoglomus:Dictyo
:cellular organisms:Bacteria:Proteobacteria:Betaproteobacteria:Methylophilales:Methylophilaceae:Methylop
:cellular organisms:Bacteria:Proteobacteria:Betaproteobacteria:Methylophilales:Methylophilaceae:Methylop
:cellular organisms:Bacteria:Proteobacteria:delta/epsilon subdivisions:Deltaproteobacteria:Desulfuromona
:cellular organisms:Bacteria:Proteobacteria:delta/epsilon subdivisions:Deltaproteobacteria:Desulfuromona
:cellular organisms:Bacteria:Proteobacteria:Alphaproteobacteria:Caulobacterales:Caulobacteraceae:Phenylo
:cellular organisms:Bacteria:Proteobacteria:Alphaproteobacteria:Caulobacterales:Caulobacteraceae:Phenylo
:cellular organisms:Bacteria:Proteobacteria:Gammaproteobacteria:Alteromonadales:Shewanellaceae:Shewanell
:cellular organisms:Bacteria:Proteobacteria:Gammaproteobacteria:Alteromonadales:Shewanellaceae:Shewanell
:cellular organisms:Bacteria:Proteobacteria:Gammaproteobacteria:Alteromonadales:Shewanellaceae:Shewanell
:cellular organisms:Bacteria:unclassified Bacteria:unclassified Bacteria (miscellaneous):halophilic euba
:cellular organisms:Bacteria:unclassified Bacteria:unclassified Bacteria (miscellaneous):halophilic euba
:cellular organisms:Bacteria:Proteobacteria:delta/epsilon subdivisions:Deltaproteobacteria:Myxococcales
:cellular organisms:Bacteria:Proteobacteria:delta/epsilon subdivisions:Deltaproteobacteria:Myxococcales
:cellular organisms:Bacteria:Proteobacteria:delta/epsilon subdivisions:Deltaproteobacteria:Myxococcales
:cellular organisms:Bacteria:Proteobacteria:delta/epsilon subdivisions:Deltaproteobacteria:Myxococcales
:cellular organisms:Bacteria:Proteobacteria:delta/epsilon subdivisions:Deltaproteobacteria:Myxococcales
:cellular organisms:Bacteria:Proteobacteria:delta/epsilon subdivisions:Deltaproteobacteria:Myxococcales
:cellular organisms:Bacteria:Proteobacteria:delta/epsilon subdivisions:Deltaproteobacteria:Myxococcales
:cellular organisms:Bacteria:Proteobacteria:delta/epsilon subdivisions:Deltaproteobacteria:Myxococcales
:cellular organisms:Bacteria:Proteobacteria:delta/epsilon subdivisions:Deltaproteobacteria:Myxococcales
:cellular organisms:Bacteria:Proteobacteria:delta/epsilon subdivisions:Deltaproteobacteria:Myxococcales
:cellular organisms:Bacteria:Proteobacteria:delta/epsilon subdivisions:Deltaproteobacteria:Myxococcales
:cellular organisms:Bacteria:Proteobacteria:delta/epsilon subdivisions:Deltaproteobacteria:Myxococcales
:cellular organisms:Bacteria:Proteobacteria:delta/epsilon subdivisions:Deltaproteobacteria:Myxococcales
:cellular organisms:Bacteria:Proteobacteria:delta/epsilon subdivisions:Deltaproteobacteria:Myxococcales
```

Figure 10.1   A partial output of the taxonomy lineage output file.

7. Using Perl, Python, or Ruby, write a script that determines the total number of phyla included in Taxonomy.dat. In the last section of the chapter, we found that the 14 phyla accounted for the list of human pathogens. What does this tell us about the likelihood that a terrestrial phyla contains any human infectious organisms?

11

DEVELOPMENTAL LINEAGE CLASSIFICATION AND TAXONOMY OF NEOPLASMS

Samuel Johnson defined a lexicographer as a "harmless drudge." The drudgery of the lexicographer's tasks is beyond dispute. In the domain of medical nomenclatures, however, the harmlessness of the lexicographer is far from certain. The misuse of medical terminology can lead to medical errors, as indicated by the U.S. Joint Commission on Accreditation of Healthcare Organization's recent ban on certain common medical abbreviations. This action was taken to reduce the occurrence of medication errors that result when nonstandard abbreviations are misinterpreted. The U.S. Institute of Medicine has advocated standardized methods for collecting codified diagnostic data as a strategy for reducing medical errors.

Because modern nomenclatures are used to annotate medical data so that clinical information can be merged with heterogeneous data sources (e.g., tissue bank records, research data sets, epidemiologic databases), the duties of lexicographers have broadened to include a range of informatics activities. For this reason, the modern curator is involved in codifying terms (providing a unique identifier to a term and all its synonyms) and mapping terms between different nomenclatures. In the past, nomenclatures were recorded on paper documents. Brevity was appreciated, and rare lesions may have been neglected. Modern nomenclatures are stored electronically. With no barriers to the size of nomenclatures, there is no reason to exclude any used terms.

As a sample implementation of a modern nomenclature, the Developmental Lineage Classification and Taxonomy of Neoplasms, hereinafter called "the neoplasm taxonomy," was used. The neoplasm taxonomy has several properties that make it particularly suitable for students:

1. It is a free, open-access medical nomenclature.
2. It has been described in the medical literature.
3. New versions of the nomenclature are made available for public download by the author at

   http://www.julesberman.info/devclass.htm.

4. It is an easily parsed XML document, with every term appearing as a lower-case alphanumeric phrase.
5. It is intended to be a comprehensive listing of all items in the knowledge domain (i.e., names of neoplasms).

The purpose of the taxonomy is to provide a listing of all names of neoplasms, with synonyms grouped under a common code number. The current version of the Neoplasm Classification contains over 135,000 unique names of neoplasms. In constructing the taxonomy, enormous effort was made to list every variant name for every known neoplasm of man. Variant names included different terms for the same concept and different ways of expressing an individual term (e.g., variations in word order).

11.1 Building the Doublet Hash

The utility of the doublet method is derived in part from the observation that most medical terms are multiword terms. In the Neoplasm Classification, all but about 250 terms are multiword terms. Unlike single words, which often have several different meanings, multiword medical terms, with very rare exceptions, have a single, specific meaning.

In Chapter 9, Section 9.2, we learned that any multiword term can be constructed by a concatenation of overlapping doublets.

For example:

Serous borderline ovarian tumor -> ("serous borderline," "borderline ovarian," "ovarian tumor")

The doublets composing the multiword terms from the neoplasm nomenclature can be combined into a list. The list of nomenclature doublets can be used to determine whether a fragment of text is composed from doublets included in the list.

We would like to build a persistent data object (see Chapter 5, Section 5.2) containing all of the doublet terms found in the Neoplasm Classification. We will use the doublet list for a variety of informatics projects featured in this book.

11.1.1 Script Algorithm

1. Create two external database objects.
2. We will tie one external database object to a dictionary object composed of key–value pairs, where the keys are the neoplasm terms in the Neoplasm Classification, and the values are the empty character ("").
3. We will tie another external database object to a dictionary object composed of key–value pairs, where the keys are the collection of word doublets from the Neoplasm Classification, and the values are the empty character ("").
4. Open the Neoplasm Classification for parsing. The compressed file is available for download at

   http://www.julesberman.info/neoclxml.gz.

   Make certain that the unzipped file is named neocl.xml and that your script lists its correct subdirectory location on your computer.

5. Parse through the file, line by line.
6. Neoplasm terms are flanked by angle brackets and can be extracted with a simple regex expression.
7. The neoplasm term is added as a new key to the dictionary object containing the terms in the nomenclature.
8. The term is parsed into doublets by iterating through each word in the term and appending the next consecutive word. Add each doublet term to the dictionary object containing word doublets as keys.
9. After the entire nomenclature file is parsed, the two dictionary objects achieve persistence through the external database objects to which they were tied.

Perl Script

```
#!/usr/local/bin/perl
use Fcntl;
use SDBM_File;
tie %doubhash, "SDBM_File", 'doub', O_RDWR|O_CREAT|O_EXCL, 0644;
tie %literalhash, "SDBM_File", 'literal', O_RDWR|O_CREAT|O_EXCL, 0644;
open (TEXT,"c\:\\ftp\\neocl.xml")||die"Cannot";
my $line = " ";
while ($line ne "")
  {
  $line = <TEXT>;
  $line =~ /\"\> ?(.+) ?\<\//;
  $phrase = $1;
  $phrase =~ s/\b([a-z]+oma)s/$1/g;
  $phrase =~ s/\b(tumo[u]?r)s/$1/g;
  $literalhash{$phrase} = "";
  @hoparray = split(/ /,$phrase);
  for ($i=0;$i<(scalar(@hoparray)+1);$i++)
      {
      if (exists $doubhash{"$hoparray[$i] $hoparray[$i+1]"})
          {
          next;
          }
      if ($hoparray[$i+1] ne "")
          {
          $doubhash{"$hoparray[$i] $hoparray[$i+1]"}= "";
          }
      }
  }
close TEXT;
untie %doubhash;
untie %literalhash;
exit;
```

Python Script

```
#!/usr/local/bin/python
import anydbm, string, re
doubhash = anydbm.open('doub', 'n')
literalhash = anydbm.open('literal', 'n')
in_file = open('c:\\ftp\\neocl.xml', "r")
singular = re.compile('omas')
england = re.compile('tumou?rs?')
phrase = ""
for line in in_file:
  neoplasm_match = re.search(r'\"\> ?(.+) ?\<', line)
  if neoplasm_match:
    phrase = neoplasm_match.group(1)
    phrase = singular.sub("oma",phrase)
    phrase = england.sub("tumor",phrase)
    literalhash[phrase] = ""
  hoparray = phrase.split()
  hoparray.append(" ")
  for i in range(len(hoparray)-1):
      doublet = hoparray[i] + " " + hoparray[i + 1]
      if doubhash.has_key(doublet):
          continue
      doubhash_match = re.search(r'[a-z]+ [a-z]+', doublet)
      if doubhash_match:
          doubhash[doublet] = ""
doubhash.close()
literalhash.close()
exit
```

Ruby Script

```
#!/usr/local/bin/ruby
require 'dbm'
mesh_file = File.open('c:/ftp/neocl.xml', 'r')
doublethash = DBM.open('doub')
literalhash = DBM.open('literal')
hoparray = Array.new()
phrase = String.new()
mesh_file.each_line do
  |line|
  line =~ /\"\> ?(.+) ?\<\//
  phrase = $1
  phrase.sub!(/\b([a-z]+oma)s/,'\1') if phrase =~ /\b([a-z]+oma)s/
  phrase.sub!(/\b(tumou?r)s/,'\1') if phrase =~ /\b(tumou?r)s/
  literalhash[phrase] = ""
  if phrase =~ /[a-z]+ [a-z]+/
    hoparray = phrase.split
  else
    next
```

```
      end
   hoparray.push(" ")
   hoparray.each_index do
         |i|
         next if doublethash.has_key? "hoparray[i] $hoparray[i+1]"
         if hoparray[i+1] =~ /[a-z]/
           doublethash[hoparray[i] + " " + hoparray[i+1]] = ""
         end
   end
end
doublethash.close
literalhash.close
exit
```

11.1.2 Analysis

We now have persistent data objects in external database files (i.e., the terms object and the doublets object) that we can use in the next section.

11.2 Scanning the Literature for Candidate Terms

Here is a simple method for extracting candidate new terms from any large corpus of text.

The method depends on the empirical observation that terms in a nomenclature are composed almost exclusively of doublets found in other terms in the same nomenclature.

The current version of the neoplasm nomenclature contains 135,000 unique terms. Of these terms, 126,756 terms are classified terms and are composed of at least two words (i.e., are doublets or greater in length). Of these 126,756 terms, all but 6,308 (4.97%) are composed entirely of doublets extracted from other terms in the reference nomenclature. This means that 95% of the classified terms from the nomenclature are formed entirely of doublet terms found in other terms from the same nomenclature.

The method compares connected word doublets in a medical text against a list of word doublets found in a nomenclature. Text phrases composed of sequences of word doublets found in an existing nomenclature are candidate new nomenclature terms. This general method can be used with any text and any existing nomenclature. This method permits curators to continually enhance their nomenclatures with new terms, an essential activity needed to ensure the proper coding and annotation of biomedical data.

11.2.1 Script Algorithm

The following algorithm parses through text, extracting candidate term phrases:

1. Collect all the doublets that occur in the entire nomenclature (i.e., use the database object created in Section 11.1).
2. Parse text (in this case individual abstract titles) into an ordered array of overlapping doublets (as per the example shown for the text string, "serous borderline ovarian tumor").

   The text file that we use is cancer_gene_titles.txt (1,752,432 bytes), created in Chapter 9, Section 9.1. It contains 28,102 titles related to the topic of genes and cancer or tumors. It is available for download at

   http://www.julesberman.info/book/cancer_gene_titles.txt.

   Alternatively, you can create your own file of titles by downloading a PubMed search on a topic of your own interest and collecting the titles, using the script provided in the previous section.
3. Compare each consecutive text doublet against the array of doublets from the nomenclature to determine whether the doublet exists somewhere in the nomenclature.
4. If the doublet from the text does not exist in the nomenclature, it can be deleted. If it exists in the nomenclature, it is concatenated with the following doublet if the following doublet exists in the nomenclature. Otherwise, it is deleted. This process continues, concatenating doublets that exist somewhere in the nomenclature. Extraneous leading words (*the, in, of, with, and*) and trailer words (*the, and, with, from, a*) are automatically deleted from the final concatenated sequence. Final concatenated sequences of two or greater consecutive doublets that match to doublets from the nomenclature are saved as candidate terms.

Perl Script

```
#!/usr/local/bin/perl
use Fcntl;
use SDBM_File;
tie %doubhash, "SDBM_File", 'doub', O_RD, 0644;
tie %literalhash, "SDBM_File", 'literal', O_RD, 0644;
open (TEXT, "c\:\\big\\cancer_gene_titles.txt")||die"cannot";
$line = " ";
$count = 0;
while ($line ne "")
   {
   $bigline = $line = <TEXT>;
   $bigline =~ s/\n//;
   $bigline =~ s/\b([a-z]+oma)s/$1/g;
   $bigline =~ s/\b(tumo[u]?r)s/$1/g;
   $englishline = "";
   @hoparray = split(/ /,$bigline);
   for ($i=0;$i<(scalar(@hoparray));$i++)
```

```perl
        {
        $doublet = "$hoparray[$i] $hoparray[$i+1]";
        if (exists $doubhash{$doublet})
            {
            if ($englishline ne "")
                {
                $englishline = $englishline . " $hoparray[$i+1]";
                }
            else
                {
                $englishline = $doublet;
                }
            }
        else
            {
            if ($englishline ne "")
                {
                $englishline =~ s/^the //o;
                $englishline =~ s/ the$//o;
                $englishline =~ s/^in //o;
                $englishline =~ s/ in$//o;
                $englishline =~ s/^of //o;
                $englishline =~ s/ of$//o;
                $englishline =~ s/^and //o;
                $englishline =~ s/ and$//o;
                $englishline =~ s/^with //o;
                $englishline =~ s/ with$//o;
                $englishline =~ s/^from //o;
                $englishline =~ s/ from$//o;
                $englishline =~ s/ a$//o;
                $englishline =~ s/^a //o;
                next if (exists $literalhash{$englishline});
                next if (exists $newhash{$englishline});
                next if ($englishline !~ / [a-z]+ /);
                $count++;
                print $count . " " . $englishline . "\n";
                $newhash{$englishline} = "";
                }
            }
        }
    }
untie %doubhash;
untie %literalhash;
exit;
```

Python Script

```python
#!/usr/local/bin/python
import anydbm, string, re
doubhash = anydbm.open('doub')
```

```
literalhash = anydbm.open('literal')
newhash = {}
in_file = open('c:\\big\\cancer_gene_titles.txt', 'r')
line = " "
count = 0
singular = re.compile('omas')
england = re.compile('tumou?rs?')
for line in in_file:
    bigline = line.rstrip(" \n")
    bigline = singular.sub("oma", bigline)
    bigline = england.sub("tumor", bigline)
    englishline = ""
    hoparray = bigline.split()
    hoparray.append(" ")
    for i in range(len(hoparray) - 1):
      doublet = hoparray[i] + " " + hoparray[i + 1]
      if doubhash.has_key(doublet):
        if (englishline != ""):
          englishline = englishline + " " + hoparray[i + 1]
        else:
          englishline = doublet
      else:
        if englishline != "":
          englishline = englishline.strip()
          englishline = re.sub(r'^the ', "", englishline)
          englishline = re.sub(r'^in ', "", englishline)
          englishline = re.sub(r'^of ', "", englishline)
          englishline = re.sub(r'^and ', "", englishline)
          englishline = re.sub(r'^with ', "", englishline)
          englishline = re.sub(r'^from ', "", englishline)
          englishline = re.sub(r'^ a', "", englishline)
          englishline = re.sub(r' the$', "", englishline)
          englishline = re.sub(r' in$', "", englishline)
          englishline = re.sub(r' of$', "", englishline)
          englishline = re.sub(r' and$', "", englishline)
          englishline = re.sub(r' with$', "", englishline)
          englishline = re.sub(r' from$', "", englishline)
          englishline = re.sub(r' a$', "", englishline)
          if literalhash.has_key(englishline):
            continue
          if newhash.has_key(englishline):
            continue
          phrase_match = re.search(r' [a-z]+ ', englishline)
          if phrase_match:
            count = count + 1
            print str(count) + " " + englishline
            newhash[englishline] = ""
doubhash.close()
```

```
literalhash.close()
exit
```

Ruby Script

```ruby
#!/usr/local/bin/ruby
require 'dbm'
in_file = File.open('c:/big/cancer_gene_titles.txt', 'r')
doublethash = DBM.open('doub')
literalhash = DBM.open('literal')
newhash = Hash.new()
hoparray = Array.new()
phrase = String.new()
count = 0
in_file.each_line do
  |line|
  line.chomp!
  line.sub!(/\b([a-z]+oma)s/,'\1') if line =~ /\b([a-z]+oma)s/
  line.sub!(/\b(tumou?r)s/,'\1') if line =~ /\b(tumou?r)s/
  englishline = String.new()
  hoparray = line.split
  hoparray.push(" ")
  hoparray.each_index do
      |i|
    next unless hoparray[i+1]
    if doublethash.has_key? hoparray[i] + " " + hoparray[i+1]
      if englishline =~ /[a-z]/
        englishline = englishline + " " + hoparray[i+1]
      else
        englishline = hoparray[i] + " " + hoparray[i+1]
      end
    else
      if englishline
        englishline.strip!
        englishline.sub!(/^the /, "") if englishline =~ /^the /
        englishline.sub!(/ the$/, "") if englishline =~ / the$/
        englishline.sub!(/^in /, "") if englishline =~ /^in /
        englishline.sub!(/ in$/, "") if englishline =~ / in$/
        englishline.sub!(/^of /, "") if englishline =~ /^of /
        englishline.sub!(/ of$/, "") if englishline =~ / of$/
        englishline.sub!(/^and /, "") if englishline =~ /^and /
        englishline.sub!(/ and$/, "") if englishline =~ / and$/
        englishline.sub!(/^with /, "") if englishline =~ /^with /
        englishline.sub!(/ with$/, "") if englishline =~ / with$/
        englishline.sub!(/^from /, "") if englishline =~ /^from /
        englishline.sub!(/ from$/, "") if englishline =~ / from$/
        englishline.sub!(/^a /, "") if englishline =~ /^a /
        englishline.sub!(/ a$/, "") if englishline =~ / a$/
        next if literalhash.has_key? englishline
```

```
            next if newhash.has_key? englishline
            next if englishline !~ / [a-z]+ /
            count = count + 1
            puts count.to_s + " " + englishline
            newhash[englishline] = ""
          end
        end
    end
end
in_file.close
doublethash.close
literalhash.close
exit
```

11.2.2 Analysis

Parsing the file cancer_genes_titles.txt, we found about 4,100 new candidate neoplasm terms. Here are some final terms from the output list:

> intraneural perineurioma of the oral mucosa
> due to promyelocytic leukemia
> spinal cord primary extragonadal
> spinal cord primary extragonadal sac tumor
> epithelioid and spindle cell haemangioma of bone
> cervical malformation neurofibromatosis type 1
> osteoblastoma of the scapula
> ameloblastic carcinoma in
> pancreatic serous cystadenoma endocrine tumor
> extrarenal rhabdoid tumor of the cervical spine
> diffuse type cell tumor of the subcutaneous
> ewing sarcoma neuroectodermal tumor of the kidney
> low grade fibromyxoid sarcoma of the colon
> inflammatory myofibroblastic tumor of the tongue
> superficial angiomyxoma the floor of the mouth
> young adult with acute lymphoblastic leukemia
> burkitt lymphoma in pediatric
> peripheral primitive neuroectodermal tumor of the maxilla
> anaplastic large cell lymphoma of bone

A cursory examination of this small portion of the 4077 returned candidate terms indicates that some of the terms seem to be legitimate names of neoplasms, which should be added to our neoplasm vocabulary:

> intraneural perineurioma of the oral mucosa
> epithelioid and spindle cell haemangioma of bone

osteoblastoma of the scapula
pancreatic serous cystadenoma endocrine tumor
extrarenal rhabdoid tumor of the cervical spine
low grade fibromyxoid sarcoma of the colon
inflammatory myofibroblastic tumor of the tongue
peripheral primitive neuroectodermal tumor of the maxilla
anaplastic large cell lymphoma of bone

The majority of terms are phrases that happen to consist of doublets from our nomenclature, but do not rise to the level of a new neoplasm term:

due to promyelocytic leukemia
spinal cord primary extragonadal
spinal cord primary extragonadal sac tumor
cervical malformation neurofibromatosis type 1
ameloblastic carcinoma in
diffuse type cell tumor of the subcutaneous
ewing sarcoma neuroectodermal tumor of the kidney
young adult with acute lymphoblastic leukemia
burkitt lymphoma in pediatric

There was one term that seems to be a poorly worded representation of a proper neoplasm's name:

superficial angiomyxoma the floor of the mouth

It should be

superficial angiomyxoma of the floor of the mouth

The original file of abstracts that contained the words *cancer* and *gene* exceeded 213 megabytes (MB) in length. The perfect curator would have read each abstract, writing down the names of neoplasms that were not contained in the nomenclature. The modern curator had the option of extracting the titles from the articles, and parsing through the titles, extracting about 4,100 candidate terms, and then examining the candidate terms to find likely new terms for the nomenclature. The semiautomated process takes about one-half hour and provides hundreds of new terms that can be added to the nomenclature.

11.3 Adding Terms to the Neoplasm Classification

One of the most common tasks in informatics is the preparation of a subtraction list (items present in one list and absent from another).

Curators need to prepare a subtraction list whenever they want to add terms to a preexisting nomenclature. The list of candidate terms must be checked against the list of terms found in the nomenclature, with removal of redundant terms in the new list.

We can use the Neoplasm Classification as a sample nomenclature. We will use the file neocl.lst (available at http://www.julesberman.info/book/neocl/lst), which contains the following list of candidate terms:

    prostate cancer
    adenocarcinoma of prostate
    spiradenocylindroma of the kidney
    spiradenocylindroma
    pleomorphic myxoid liposarcoma
    spindle cell myxoid liposarcoma
    matrix producing carcinoma of breast
    matrix producing carcinoma of the breast
    dini of breast
    precancer flat epithelial atypia
    matrix-producing carcinoma of the breast
    early onset cancer
    early-onset neoplasm
    early-onset neoplasia
    carcinoma of the bellini collecting duct
    adenocarcinoma of the prostate

We need to know which terms, among the candidate terms, are already included in the Neoplasm Classification.

11.3.1 Script Algorithm

1. Open the Neoplasm Classification file.
2. Parse through the file, collecting every code/term pair in the Neoplasm Classification, and assigning each pair as the key and value (respectively) for a dictionary object.
3. Open the file containing the list of candidate terms to be added to the Neoplasm Classification.
4. Parse each term from the list, checking to see if it is already contained as a key in the dictionary object.
5. For each term, if the term does not already exist as a key in the dictionary object, print it to an external file.
6. After the script executes, you have a new file containing terms that can be added to the Neoplasm Classification.

Perl Script

```perl
#!/usr/local/bin/perl
open (TEXT,"c\:\\ftp\\neocl.xml")||die"Cannot";
$line = " ";
while ($line ne "")
   {
   $line = <TEXT>;
   next if ($line !~ /C[0-9]{7}/);
   $line =~ /\"\> ?(.+) ?\<\//;
   $phrase = $1;
   $doubhash{$phrase}="";
   }
close TEXT;
open (TEXT,"c\:\\ftp\\neocl.lst")||die"Cannot";
open (OUT,">new.out")||die"Cannot";
$line = " ";
while ($line ne "")
   {
   $line = <TEXT>;
   $line =~ s/\n//o;
   next if ($line eq "");
   if (exists $doubhash{$line})
       {
       print "$line already exists\n";
       }
   clsc
       {
       print OUT "$line\n";
       }
   }
exit;
```

Python Script

```python
#!/usr/local/bin/python
import re, string
vocab_in = open("c:\\ftp\\neocl.xml", "r")
doub_hash = {}
for line in vocab_in:
   code_match = re.search(r'C[0-9]{7}', line)
   if not code_match:
      continue
   line_match = re.search(r'\"\> ?(.+) ?\<\/', line)
   if line_match:
      phrase = line_match.group(1)
      doub_hash[phrase] = ""
vocab_in.close()
```

```
candidate_file = open("c:\\ftp\\neocl.lst", "r")
out_file = open("new.out", "w")
for line in candidate_file:
  line = re.sub(r'\n',"", line)
  if (line == ""):
    continue
  if doub_hash.has_key(line):
    print line + " already exists"
  else:
    print>>out_file, line
exit
```

Ruby Script

```
#!/usr/local/bin/ruby
vocab_in = File.open("c:/ftp/neocl.xml", "r")
doub_hash = {}
vocab_in.each_line do
  |line|
  next if (line !~ /C[0-9]{7}/)
  line =~ /\"\> ?(.+) ?\<\//
  phrase = $1
  doub_hash[phrase] = ""
end
vocab_in.close
candidate_file = File.open("c:/ftp/neocl.lst", "r")
out_file = File.open("new.out", "w")
candidate_file.each_line do
  |line|
  line.sub!(/\n/,"")
  next if (line == "")
  if doub_hash.has_key? line
    puts line + " already exists"
  else
    out_file.puts line
  end
end
exit
```

11.3.2 Analysis

The script splits the output into the set of terms already contained in the Neoplasm Classification, displayed on the computer monitor:

> prostate cancer already exists
> adenocarcinoma of prostate already exists
> spiradenocylindroma of the kidney already exists
> matrix producing carcinoma of breast already exists

matrix producing carcinoma of the breast already exists
matrix-producing carcinoma of the breast already exists
adenocarcinoma of the prostate already exists

And an output file, containing the list of terms that are not already included in the Neoplasm Classification:

spiradenocylindroma
pleomorphic myxoid liposarcoma
spindle cell myxoid liposarcoma
dini of breast
precancer-flat epithelial atypia
early onset cancer
early-onset neoplasm
early-onset neoplasia
carcinoma of the bellini collecting duct

11.4 Determining the Lineage of Every Neoplasm Concept

Biological classifications drive down the complexity of nomenclatures by assigning every term to a class of objects that contain similar features, inherited from a lineage of ancestral objects. We have seen, in the prior chapter, that knowing the lineages of organisms can lead to treatments for newly encountered pathogens. Similarly, knowing the lineage of neoplasms may help us find the tumors most likely to respond, as a biological class, to molecular-targeted cancer treatments. The importance of tumor lineage is one of the important concepts discussed in my book, *Neoplasms: Principles of Development and Diversity* (Jones & Bartlett Publishers, 2009).

The Neoplasm Classification contains about 135,000 names of neoplasms, organized under about 6,000 concepts. A *concept* is the collection of synonyms for a specific type of neoplasm. Every neoplasm term and concept can be assigned a unique position within a simple class hierarchy, consisting of several dozen ancestral classes (Figure 11.1).

The Neoplasm Classification is packaged as an XML (eXtensible Markup Language) file. The terms in the nomenclature are marked up with tags that provide each term with a code number describing each term. Each term in the Neoplasm Classification is nested under another element that names a class of neoplasms. Each named class of neoplasms is nested under elements for the father of the class, and this nesting continues up the classification hierarchy.

XML is a markup language created for the Internet, and data that is delivered in XML files permits us to search for related information located anywhere in the Internet. In Chapter 18, we will be describing XML in much more detail. For now, we will take advantage of language-specific modules designed to parse XML, and we

Figure 11.1 Schematic drawing of the class structure of the Neoplasm Classification.

will determine the full neoplasm lineage for every term contained in the Neoplasm Classification. If you are unfamiliar with XML, you can skip this section of the chapter and come back to it after reading Chapter 18.

11.4.1 Script Algorithm

1. Call the XML parser module into your script.
2. Define subroutines that process XML information for specific events that occur as the XML file is parsed. These events happen whenever the script encounters the start of an XML element; the script encounters the end of an XML element; and the script encounters the data described by the XML tag.
3. Provide the parser object with the name of the XML file that you would like to parse. In this case, it is the neocl.xml file.
4. When an element is encountered, the parser passes the name of the element and any attributes within the element (in this case, the code number for the term) to a list of variables. When the data contents of the element are encountered, the parser passes the data to a variable. In this case, the data associated with an element is the neoplasm term.
5. As the parser works its way down the hierarchy, it concatenates the names of the ancestors into a string. When it finally encounters the lowest element in

the hierarchy, it concatenates the data (the name of the term), and the attribute (the code for the term), appends the hierarchical list of elements (ancestors) to the string, and prints it to an external file. When it backs up through the hierarchy (when it moves through different class lineages), it truncates the previously built string of concatenated classes to exclude nonancestral classes.

Perl Script

```perl
#!/usr/local/bin/perl
open (STDOUT, ">neoself")||die"Cannot";
my ($text, $lastname, $name, $count);
use XML::Parser;
my $parser = XML::Parser->new( Handlers => {
  Start => \&handle_elem_start,
  End => \&handle_elem_end,
  Char => \&handle_char_data,
  });
$file = "neocl.xml";
$parser -> parsefile($file);
sub handle_elem_start
  {
  ($expat, $name, %atts) = @_;
  if (exists $atts{"nci-code"})
      {
      $code = $atts{"nci-code"};
      }
  else
      {
      $lastname = $name . "\>" . $lastname;
      }
  }
sub handle_elem_end
  {
  ($expat, $name) = @_;
  if ($name eq "name")
      {
      $count++;
      $text =~ s/\n//g;
      print $count . "\|" . $text . "\|" . $code . "\|" . $lastname .
"\n";
      $text = "";
      }
  $lastname =~ s/${name}\>//g;
  }
sub handle_char_data
  {
  ($expat, $characters) = @_;
  $text = $text . $characters;
```

```
    }
exit;
```

Python Script

```python
#!/usr/local/bin/python
import xml.parsers.expat
import re
parsefile = open('c:\\ftp\\neocl.xml','r')
filestring = parsefile.read()
lastname = ""
code = ""
count = 0
text = ""
def start_element(name, attrs):
    global lastname
    global code
    if attrs.has_key("nci-code"):
      code = attrs["nci-code"]
    else:
      lastname = name + ">" + lastname
def end_element(name):
    global count
    global code
    global text
    global lastname
    if name == "name":
      count = count + 1
      print str(count) + "|" + text + "|" + code + "|" + lastname + ⤷
"\n"
      text = ""
    lastname = re.sub(name + r'>','', lastname)
def char_data(data):
    global text
    text = repr(data)
    textmatch = re.search(r'\'(.+)\'',text)
    if textmatch:
      text = textmatch.group(1)
p = xml.parsers.expat.ParserCreate()
p.StartElementHandler = start_element
p.EndElementHandler = end_element
p.CharacterDataHandler = char_data
p.Parse(filestring)
exit
```

Ruby Script

```ruby
#!/usr/local/bin/ruby
require 'rexml/document'
require 'rexml/streamlistener'
include REXML
```

```
class Listener
  include StreamListener
  @@out = File.open("neoruby.txt","w")
  @@count = 0
  @lastname = ""
  @neoplasm_name = ""
  @code = ""
  def tag_start(name, attributes)
      @code = "#{attributes}"
      @code = $& if (@code =~ /[0-9]{7}/)
      @lastname = "#{name}\>#{@lastname}" if (@code !~ /[0-9]{7}/)
  end
  def text(text)
      @neoplasm_name = text
  end
  def tag_end(name)
    @@count = @@count+1
    @@out.puts("#{@@count}\|#{@neoplasm_name}\|#{@code}\|#{@
lastname}") if (name =~/name/)
    @lastname.gsub!(/#{name}\>/, "") #if (@lastname =~ /#{name}\>/)
  end
end
listener = Listener.new
parser = Parsers::StreamParser.new(File.new("neocl.xml"), listener)
parser.parse
exit
```

11.4.2 Analysis

The neocl.xml file is over 10 MB in length. It takes several seconds, on most computers (with 2–3 GHz CPUs) to run this script, producing an output file that exceeds 17 MB in length. Here are a few lines of the output file:

1|teratoma|C3403000|totipotent_or_multipotent_differentiating>
primitive_differentiating>primitive>embryonic>neoplasms>
tumor_classification>

2|embryonal ca|C3752000|totipotent_or_multipotent_differentiating>
primitive_differentiating>primitive>embryonic>neoplasms>
tumor_classification>

3|embryonal cancer|C3752000|totipotent_or_multipotent_differentiating>
primitive_differentiating>primitive>embryonic>neoplasms>
tumor_classification>

4|embryonal carcinoma|C3752000|totipotent_or_multipotent_differentiating>
primitive_differentiating>primitive>embryonic>neoplasms>
tumor_classification>

Because we know the structure of the Neoplasm Classification file, we could have written a parsing script without using an external XML parser, if we had so chosen. The script would have been similar to the script that we used to find the lineage of organisms from the Taxonomy.dat file (Chapter 10). However, because the neocl.xml file is created as an XML file, it is better to use the readily available XML parsing module. Doing so shortens our script and, if you do much work with XML, is easy to read. Once you have learned to parse XML files, you will be able to write scripts that collect, transform, and analyze data from multiple, different XML files, collected from anywhere on the Internet.

Exercises

1. Parse through the Neoplasms Classification to determine the total number of concepts and terms in the nomenclature. Neoplasm Classification in XML format can be obtained at

   http://www.julesberman.info/neoclxml.gz.

2. Using the Neoplasms Classification in XML format, verify that no term in the nomenclature occurs more than once in the nomenclature (i.e., verify term uniqueness).

3. Using the Neoplasms Classification in XML format, verify that no concept in the nomenclature occurs more than once in the nomenclature (i.e., concept uniqueness).

4. It is an interesting fact that every proper term (a term composed of one or more words) in the Neoplasm Classification contains an "o", an "a", and an "e". The only entries in the Neoplasm Classification that lack one or more of these three letters are nonword abbreviations. These abbreviations are mmgct, mgct, itgcn, itgcnu, igcnu, xp, scc, cis, bcc, sil, hsil, hgsil, dcis, ipmt, hlrcc, bnct, ncmh, dsrct, gist, ptgc, cgl, cml, sctcl, idl, lphd, nlphd, dlbcl, upsc, cin, sspc, jgct, gtni, mnti, mpnst, pstt, punlmp, vin. In Perl, Python, or Ruby, write a script that tests every term in the Neoplasm Classification, ensuring that it contains an "o", an "a", and an "e".

5. In Chapter 8, we developed a script that extracted, from OMIM, all records that contained a specific "Oncology" section, listing the neoplastic conditions associated with a specific genetic disorder. Many records in OMIM are associated with neoplastic terms but lack an "Oncology" section. Using Perl, Python, or Ruby, write a new script that extracts all of the OMIM records that contain any of the neoplastic terms contained in the Neoplasm Classification.

6. Using Perl, Python, or Ruby, modify the script from Exercise 5 to extract the OMIM record number (of each OMIM record that containing a neoplasm term) followed by the list of neoplastic terms that are present in the OMIM record.

12
U.S. Census Files

The Census 2000 Modified Race Data Summary File (MR(31)-CO.txt) contains population data for U.S. states and counties and Puerto Rico. Data is stratified for 31 categorized races and ethnicities (Figure 12.1).

The MR(31)-CO.txt is a public domain file, about 65 megabytes (MB) in length (Figure 12.2), available from the U.S. Census Bureau at:

http://www.census.gov/popest/archives/files/MR-CO.txt

A Web page providing some background information on this file is available at

http://www.census.gov/popest/archives/files/MRSF-01-US1.html

And a data dictionary for the file is available at

http://www.census.gov/popest/archives/files/MRSF-01-US1.pdf

12.1 Total Population of the United States

One of the easiest uses of the MR(31)-CO.txt file is to count the population of the United States, or of any collection of states, or counties within states. The file lists population data for states, stratified by age, and then breaks down the populations of states by their counties (Figure 12.3).

The dictionary key for the first 8 bytes of each record is shown in Figure 12.4.

If we look at just the first record (top line of Figure 12.3), we see that the record applies to state "01", which happens to be Alabama, that it applies to the entire state (i.e., is not restricted to a county), and that the data applies to age group "1" (infants under 1 year of age). Bytes 9 and above contain population counts for 31 races and ethnicities. Line 20 marks the beginning of records for specific counties.

12.1.1 Script Algorithm

1. Open the 65+ MB MR(31)-CO.txt file.
2. Parse through each line (record) of the file, ignoring lines that contain county populations. (*Note:* The state population records lack a county code in bytes 3 to 5, and thus byte 3 is a space in the noncounty records.)

1. White alone
2. Black or African American alone
3. American Indian and Alaska Native alone
4. Asian alone
5. Native Hawaiian and Other Pacific Islander alone
6. White and Black or African American
7. White and American Indian and Alaska Native
8. White and Asian
9. White and Native Hawaiian and Other Pacific Islander
10. Black or African American and American Indian and Alaska Native
11. Black or African American and Asian
12. Black or African American and Native Hawaiian and Other Pacific Islander
13. American Indian and Alaska Native and Asian
14. American Indian and Alaska Native and Native Hawaiian and Other Pacific Islander
15. Asian and Native Hawaiian and Other Pacific Islander
16. White and Black or African American and American Indian and Alaska Native
17. White and Black or African American and Asian
18. White and Black or African American and Native Hawaiian and Other Pacific Islander
19. White and American Indian and Alaska Native and Asian
20. White and American Indian and Alaska Native and Native Hawaiian and Other Pacific Islander
21. White and Asian and Native Hawaiian and Other Pacific Islander
22. Black or African American and American Indian and Alaska Native and Asian
23. Black or African American and American Indian and Alaska Native and Asian and Native Hawaiian and Other Pacific Islander
24. Black or African American and Asian and Native Hawaiian and Other Pacific Islander
25. American Indian and Alaska Native and Asian and Native Hawaiian and Other Pacific Islander
26. White and Black or African American and American Indian and Alaska Native and Asian
27. White and Black or African American and American Indian and Alaska Native and Native Hawaiian and Other Pacific Islander
28. White and Black or African American and Asian and Native Hawaiian and Other Pacific Islander
29. White and American Indian and Alaska Native and Asian and Native Hawaiian and Other Pacific Islander
30. Black or African American and American Indian and Alaska Native and Asian and Native Hawaiian and Other Pacific Islander
31. White and Black or African American and American Indian and Alaska Native and Asian and Native Hawaiian and Other Pacific Islander

Figure 12.1   Ethnicity data dictionary for Census 2000 Modified Race Data Summary File.

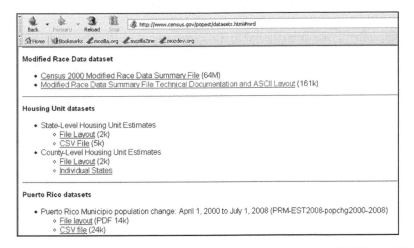

Figure 12.2   Download Page for the Census 2000 Modified Race Data Summary File.

```
                                          http://www.census.gov/popest/archives/files/MR-CO.txt
     Home   Bookmarks  mozilla.org  mozillaZine  mozdev.org

01     1     866     98    15    9     6     9     3     5     4     1     2     1     2     0
01     2    2961    314    47   23    17    19    17     9     4     4     2     2     2     1
01     3    3029    400    60   21    25    15    43    14     1     3     5     0     3     1
01     4    2378    354    48   19    13    22    26     9     1     4     4     2     0     1
01     5    3599    354    72   30    23    16    30     8     2     6     6     2     5     2
01     6    5756    311   112   52    67    16    15     6     1     2     9     0     0     1
01     7    5189    305    91   31    40     7    15     5     4     1     2     3     1     0
01     8    3866    257    70   32    32     8    11     4     0     1     0     2     4     0
01     9    2919    230    58   34    16     6    15     5     3     5     1     1     1     2
01    10    2142    201    41   16    14     8    19    11     0     3     2     2     0     0
01    11    1524    161    31   10    15     2    13     6     0     3     0     1     0     1
01    12    1079    150    13    5     8     4     4     2     0     1     5     0     1     0
01    13     708     88    12    8     6     1     7     3     1     0     3     0     0     0
01    14     508     60     8    3     3     1     1     2     0     1     2     0     0     0
01    15     415     70     5    7     5     0     3     2     0     1     0     0     0     0
01    16     282     48     1    6     2     0     1     0     1     1     2     0     0     0
01    17     209     37     2    2     3     0     2     0     0     0     4     0     0     0
01    18     120     18     0    0     0     0     1     1     1     1     0     1     0     0
01    19      91     17     4    1     2     0     0     1     0     1     1     0     0     0
01001  1       2      0     0    0     0     0     0     0     0     0     0     0     0     0
01001  2      18      0     0    0     0     0     0     0     0     0     0     0     0     0
01001  3      18      3     1    1     0     0     1     0     0     0     0     0     0     0
01001  4      27      7     0    0     0     0     0     0     0     0     0     0     0     0
01001  5      29      3     3    0     0     0     1     0     0     0     0     0     0     0
01001  6      45      0     1    0     0     0     0     0     0     0     0     0     0     0
01001  7      34      0     1    0     0     0     0     0     0     0     0     0     0     0
01001  8      24      1     1    0     0     0     0     0     0     0     0     0     0     0
01001  9      27      3     0    1     0     0     0     0     0     0     0     0     0     0
01001 10      23      2     0    0     0     0     0     0     0     0     0     0     0     0
```

Figure 12.3   The first few records in the Census 2000 Modified Race Data Summary File.

3. Extract bytes 9 to 993 of each record.
4. Add all the number entries within the byte 9 to 993 character string.
5. Add the sum to the population tally.
6. After the entire file is parsed, the population tally represents the total population of states and Puerto Rico.
7. Print the final tally to the monitor.

```
ASCII FILE LAYOUT

Character                    Description
  1–2                        FIPS state code
  3–5                        FIPS county code
  6                          Blank
  7–8                        Five–year age groups
                               1 = age 0
                               2 = ages 1–4
                               3 = ages 5–9
                               4 = ages 10–14
                               5 = ages 15–19
                               6 = ages 20–24
                               7 = ages 25–29
                               8 = ages 30–34
                               9 = ages 35–39
                              10 = ages 40–44
                              11 = ages 45–49
                              12 = ages 50–54
                              13 = ages 55–59
                              14 = ages 60–64
                              15 = ages 65–69
                              16 = ages 70–74
                              17 = ages 75–79
                              18 = ages 80–84
                              19 = ages 85+
```

Figure 12.4   Data dictionary, from file MRSF-01-US1.pdf, covering the first 8 bytes of the Census 2000 Modified Race Data Summary File.

Perl Script

```perl
#!/usr/local/bin/perl
open (TEXT, "c\:\\big\\mr\(31\)\-co\.txt")||die"cannot";
$line = " ";
$total = 0;
while ($line ne "")
    {
    $line = <TEXT>;
    next if (substr($line,2,1) !~ / /);
    $amount = substr($line, 8, 992);
    @lineitems = split(/ +/,$amount);
    $subtotal = 0;
    foreach $item (@lineitems)
       {
       $subtotal = $subtotal + $item;
       }
```

```
    $total = $total + $subtotal;
    }
print "The total US population and Puerto Rico is $total\n";
exit;
```

Python Script

```python
#!/usr/local/bin/python
import re, string
from decimal import Decimal
census_file = open('c:\\big\\mr(31)-co.txt', "r")
total = 0
for line in census_file:
  if line[2:3] == " ":
    amount = line[8:1000]
    lineitems = re.split(r' +',amount)
    for item in lineitems:
      if item.isdigit():
        total = total + int(item)
print "The total US population and Puerto Rico is " + str(total)
exit
```

Ruby Script

```ruby
#!/usr/local/bin/ruby
census_file = File.open("c:\\big\\mr\(31\)\-co\.txt")
total = 0
census_file.each do
   |line|
  next if (line.slice(2,1) !~ / /)
  lineitems = line.slice(8,992).split(/ +/)
  subtotal = 0
  lineitems.each {|item| subtotal = subtotal + item.to_i}
  total = total + subtotal
end
puts "The total US population and Puerto Rico is #{total}";
exit
```

12.1.2 Analysis

The script produces the following output:

"The total US population and Puerto Rico is 285230516"

With a slight modification of the script, you can determine the year 2000 population for any state or county, by gender, age, or ethnic group. These data are used frequently to represent medical data as a percentage of the year 2000 subpopulations or to adjust rate data collected in any year against the standard year 2000 population.

12.2 Stratified Distribution for the U.S. Census

The MR(31)-CO.txt file can be used to determine the year 2000 population, for each age category.

12.2.1 Script Algorithm

1. Open the 65+ MB MR(31)-CO.txt file.
2. Parse through each line (record) of the file, ignoring lines that contain county populations. (*Note:* The state population records lack a county code in bytes 3 to 5, and thus byte 3 is a space in the noncounty records.)
4. Extract bytes 7 and 8 from the record. These two bytes contain the age category for the record.
3. Extract bytes 9 to 993 of each record (containing the population counts for the race categories).
4. Add all the number entries within the byte 9 to 993 character string.
5. Create a dictionary object, with each of the 19 age categories as a different key for the dictionary object, and with the sum of the population categories for the age as the dictionary value. Every time a record is encountered, add the value of the records population tally to the dictionary value for its age category (key).
6. After the entire file is parsed, the dictionary object holds the total population for each of the age categories (keys).
7. Sort the dictionary keys, and print out the key and the value for all the key–value pairs of the dictionary object.

Perl Script

```perl
#!/usr/local/bin/perl
open (TEXT, "c\:\\big\\mr\(31\)\-co\.txt")||die"cannot";
$line = " ";
$total = 0;
while ($line ne "")
    {
    $line = <TEXT>;
    next if (substr($line,2,1) !~ / /);
    $age = substr($line,6,2);
    $age =~ s/ /0/;
    $amount = substr($line, 8, 992);
    @lineitems = split(/ +/,$amount);
    $subtotal = 0;
    foreach $item (@lineitems)
        {
        $subtotal = $subtotal + $item;
        }
```

```
    $agepop{$age} = $agepop{$age} + $subtotal;
    $total = $total + $subtotal;
    }
print "The total US population and Puerto Rico is $total\n";
@keysarray = sort(keys(%agepop));
foreach $key (@keysarray)
    {
    print "$key $agepop{$key}\n";
    }
exit;
```

Python Script

```
#!/usr/local/bin/python
import re, string
from decimal import Decimal
census_file = open('c:\\big\\mr(31)-co.txt', "r")
total = 0
agepop = {}
for line in census_file:
  if line[2:3] == " ":
    amount = line[8:1000]
    age = line[6:8]
    age = age.lstrip()
    lineitems = re.split(r' +',amount)
    subtotal = 0
    for item in lineitems:
      if item.isdigit():
          subtotal = subtotal + int(item)
    if agepop.has_key(age):
      agepop[age] = int(agepop[age]) + subtotal
    else:
      agepop[age] = subtotal
    total = total + subtotal
print "The total US population and Puerto Rico is " + str(total)
keylist = agepop.keys()
keylist = [int(x) for x in keylist]
keylist = sorted(keylist)
keylist = [str(x) for x in keylist]
for key in keylist:
  print key, agepop[key]
exit
```

Ruby Script

```
#!/usr/local/bin/ruby
census_file = File.open("c:\\big\\mr\(31\)\-co\.txt")
total = 0
agepop = Hash.new(0)
```

```
census_file.each do
   |line|
   next if (line.slice(2,1) !~ / /)
   age = line.slice(6,2)
   age.sub!(/ /,"0") if age.include? " "
   lineitems = line.slice(8,992).split(/ +/)
   subtotal = 0
   lineitems.each {|item| subtotal = subtotal + item.to_i}
   agepop[age] = agepop[age] + subtotal
   total = total + subtotal
end
puts "The total US population and Puerto Rico is #{total}"
agepop.keys.sort.each {|k| print k, " ", agepop[k], "\n"}
exit
```

12.2.2 Analysis

Here is the output of the script:

```
The total US population and Puerto Rico is 285230516
01 3863691
02 15607513
03 20854667
04 20833872
05 20533326
06 19265192
07 19652843
08 20773213
09 22971513
10 22692677
11 20325524
12 17815464
13 13658120
14 10966011
15 9667826
16 8964111
17 7498891
18 4998769
19 4287293
```

The top line states the total population. The next 19 lines provide the population for each of the 19 age groups.

12.3 Adjusting for Age

Suppose you are studying disease rates of whooping cough (a disease of childhood) in two populations. If the first population has a large proportion of children, then it will likely have a higher incidence of whooping cough in its population compared with another population with a low proportion of children, because the "at-risk" population is higher. To determine whether the first population has a true, increased rate of whooping cough, we need to somehow adjust for the differences in the proportion of young people in the two populations.

An age-adjusted rate is the crude rates of cancer in an age category, weighted against the proportion of persons in the age groups of a standard population. When we age-adjust rates, we cancel out the changes in the rates of disease occurrence in different populations that result from differences in the proportion of people in different age groups.

The NCI SEER program provides a step-by-step example demonstrating how it age-adjusts population-based cancer data (Figure 12.5), available at

http://seer.cancer.gov/seerstat/tutorials/aarates/step3.html

Let us use SEER's data to recompute the age-adjusted cancer incidence rate for the United States.

Here is a data abstraction of the SEER data sample. Each line record consists of the age group, the number of cancer cases in the age group, the sample population size, and the

Age	Count	Population	Crude Rate	US 2000 Standard Populations	Age Distribution of Std Pop	
00 years	29	139,879	20.7	3,794,901	0.013818	0.29
01-04 years	87	553,189	15.7	15,191,619	0.055316	0.87
05-09 years	67	736,212	9.1	19,919,840	0.072532	0.66
10-14 years	71	770,999	9.2	20,056,779	0.073031	0.67
15-19 years	87	651,390	13.4	19,819,518	0.072167	0.96
20-24 years	177	639,159	27.7	18,257,225	0.066478	1.84
25-29 years	290	676,354	42.9	17,722,067	0.064530	2.77
30-34 years	657	736,557	89.2	19,511,370	0.071045	6.34
35-39 years	1,072	724,826	147.9	22,179,956	0.080762	11.94
40-44 years	1,691	700,200	241.5	22,479,229	0.081852	19.77
45-49 years	2,428	617,437	393.2	19,805,793	0.072117	28.36
50-54 years	2,931	516,541	567.4	17,224,359	0.062718	35.59
55-59 years	2,881	361,170	797.7	13,307,234	0.048454	38.65
60-64 years	2,817	259,440	1,085.8	10,654,272	0.038794	42.12
65-69 years	2,817	206,204	1,366.1	9,409,940	0.034264	46.81
70-74 years	2,744	172,087	1,594.5	8,725,574	0.031772	50.66
75-79 years	2,634	142,958	1,842.5	7,414,559	0.026998	49.74
80-84 years	1,884	99,654	1,890.5	4,900,234	0.017843	33.73
85+ years	1,705	92,692	1,839.4	4,259,173	0.015509	28.53
All Ages				274,633,642	1.000000	400.3

Figure 12.5 Figure of a sample age-adjusted data set, produced by SEER.

total population of the age group, in the United States, for a selected standard year. Each entry of each line record is separated by a comma (so-called CSV, or comma-separated value format). You can check that these are the same numbers provided in the SEER sample. With these numbers, we can calculate all of the columns found in the SEER chart:

```
0,29,139879,3794901
1,87,553189,15191619
5,87,736212,19919840
10,71,770999,20056779
15,87,651390,19819518
20,177,639159,18257225
25,290,676354,17722067
30,657,736557,19511370
35,1072,724826,22179956
40,1691,700200,22479229
45,2428,617437,19805793
50,2931,516541,17224359
55,2881,361170,13307234
60,2817,259440,10654272
65,2817,206204,9409940
70,2744,172087,8725574
75,2634,142958,7414559
80,1884,99654,4900234
85,1705,92692,4259173
```

Note that the last column, the population data for each age category, provided by SEER for this example, is very close to the data we calculated in the prior section, from the entire MR(31)-CO.txt file.

The age-adjusted cancer rate for the sample population requires a succession of very simple calculations.

12.3.1 Script Algorithm

1. Load the comma-separated data (above) into an external file, which you can name pop_data.txt. The data.txt file is available at http://www.julesberman.info/ book/pop_data.txt.
2. Open the file, and parse each of the lines of the data file, splitting the individual comma-separated items of each line into an array object.
3. Calculate the age-specific crude cancer incidence (the number of cases divided by the size of the age-specific population in the studied community), and create a dictionary object with the age as the key, and the crude incidence as the value. Create an incrementing tally of the standard U.S. population by adding item three of each line array (the population of the age group in the United States).

4. After the data file has been parsed, we are left with a dictionary object containing all of the crude cancer incidence rates for each age group, and a variable containing the total population of the United States.

5. Close and reopen the data file. This will reset the file to its beginning.

6. Parse through the line arrays of the data file once more. For each data line, calculate the weight of the age-group population (the fraction of the total U.S. population for the particular age group), and the weighted cancer incidence rate for the age group (the crude age-specific cancer incidence rate multiplied by the age-group weight).

7. As you parse through the data file, keep an incremental tally of the age-adjusted cancer incidence rate for the total population by adding each new age-weighted cancer incidence rate.

8. Print out a summary list, with all of the columns and entries of the SEER file.

9. Print out the total age-adjusted cancer incidence rate for the U.S. population, and the total population of the standard U.S. population applied in the exercise.

Perl Script

```perl
#!/usr/local/bin/perl
open (CSV, "c\:\\ftp\\pop_data.txt")||die"cannot";
open (OUT, ">pop_data.out");
$linc = " ";
$total_age = 0;
while($line ne "")
   {
   $line = <CSV>;
   @age_array = split(/\,/,$line);
   if ($age_array[2] != 0)
      {
      $crude_rate =  ($age_array[1] / $age_array[2]) *  100000;
      $crudc_dict{$age_array[0]} = $crude_rate;
      $total_age = $total_age + $age_array[3]
      }
   }
close CSV;
open (CSV, "c\:\\ftp\\pop_data.txt")||die"cannot";
$line = " ";
$total_rate = 0;
while($line ne "")
   {
   $line = <CSV>;
   @age_array = split(/\,/,$line);
   if ($age_array[2] != 0)
      {
      $weight =  ($age_array[3] / $total_age);
      $age_rate = $weight * $crude_dict{$age_array[0]};
      $total_rate = $total_rate + $age_rate;
```

```perl
    printf OUT ("%-3.3d %-6.6d %-8.8d %-10.04f %-10.10d %-2.04f
%-6.03f\n",
    $age_array[0], $age_array[1], $age_array[2],
    $crude_dict{$age_array[0]}, $age_array[3], $weight, $age_rate);
    }
  }
print OUT "\nThe age-adjusted population cancer rate is " .
$total_rate . "\n";
print OUT "The total population is " . $total_age . "\n";
exit;
```

Python Script

```python
#!/usr/local/bin/python
import re, string
data_in = open("c:/ftp/pop_data.txt", "r")
data_out = open("pop_data.out", "w")
age_array = []
crude_dict = {}
total_age = 0
for line in data_in:
  line = string.rstrip(line)
  age_array = re.split(r',', line)
  if (age_array[2]):
    crude_rate =  (float(age_array[1]) / float(age_array[2])) *
100000
    crude_dict[age_array[0]] = crude_rate
    total_age = total_age + int(age_array[3])
data_in.close()
data_in = open("c:/ftp/pop_data.txt")
total_rate = 0;
for line in data_in:
  line = string.rstrip(line)
  age_array = re.split(r',', line)
  if (age_array[2]):
    weight =  (float(age_array[3]) / total_age)
    age_rate = weight * float(crude_dict[age_array[0]])
    total_rate = total_rate + age_rate
    print_format = "%-3.3d %-6.6d %-8.8d %-10.04f %-10.10d %-2.04f
%-6.03f"
    print>>data_out, print_format % (int(age_array[0]),
int(age_array[1]),
 int(age_array[2]), float(crude_dict[age_array[0]]),
int(age_array[3]),
 float(weight), float(age_rate))
print>>data_out, "\nThe age-adjusted population cancer rate is " +
str(total_rate)
print>>data_out, "The total population is " + str(total_age)
exit
```

Ruby Script

```
#!/usr/local/bin/ruby
require 'mathn'
data_in = File.open("c:/ftp/pop_data.txt", "r")
data_out = File.open("pop_data.out", "w")
age_array = []
crude_dict = {}
total_age = 0
data_in.each_line do
  |line|
  line.chomp!
  age_array = line.split(/\,/)
  if (age_array[2])
    crude_rate =  (age_array[1].to_i / age_array[2].to_i).to_f *
100000
    crude_dict[age_array[0]] = crude_rate
    total_age = total_age + age_array[3].to_i
  end
end
data_in.close()
data_in = File.open("c:/ftp/pop_data.txt")
total_rate = 0;
data_in.each_line do
  |line|
  line.chomp!
  age_array = line.split(/\,/)
  if (age_array[2])
    weight =  (age_array[3].to_i / total_age).to_f
    age_rate = weight * crude_dict[age_array[0]]
    total_rate = total_rate + age_rate
    data_out.printf("%-3.3d %-6.6d %-8.8d %-10.04f %-10.10d %-2.04f
%-6.03f\n",
    age_array[0], age_array[1], age_array[2], crude_dict[age_
array[0]],
    age_array[3], weight, age_rate)
  end
end
data_out.puts "\nThe age-adjusted population cancer rate is " +
total_rate.to_s
data_out.puts  "The total population is " + total_age.to_s
exit
```

12.3.2 Analysis

The output is shown in Figure 12.6.

Check this output against the official SEER data we reviewed at the beginning of this section. The calculated data are identical.

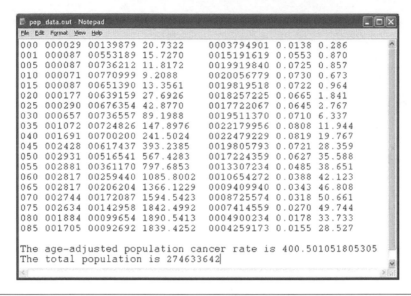

Figure 12.6   Output of the age-adjusted total population cancer incidence rate.

Exercises

1. In Section 12.2, we stratified the U.S. population into 19 age bins. Using Perl, Python, or Ruby, modify the script so that the first column in the output is the age range included in each bin number, rather than the bin number itself.

2. Using Perl, Python, or Ruby, determine the most populous county for each state.

3. Using Perl, Python, or Ruby, determine the state with the youngest population and the state with the oldest population (by average age).

4. A summary table of U.S. population and racial subpopulations (Figure 12.7) can be found in the data dictionary file, at

   http://www.census.gov/popest/archives/files/MRSF-01-US1.pdf

Table 1. Summary of Modified Race and Census 2000 Race Distributions for the United States				
Subject	Modified Race		Census 2000	
	Number	Percent	Number	Percent
TOTAL POPULATION	281,421,906	100.00	281,421,906	100.00
One race	277,524,226	98.62	274,595,678	97.57
Specified race only	277,524,226	98.62	259,236,605	92.12
White	228,104,485	81.05	211,460,626	75.14
Black or African American	35,704,124	12.69	34,658,190	12.32
American Indian and Alaska Native	2,663,818	0.95	2,475,956	0.88
Asian	10,589,265	3.76	10,242,998	3.64
Native Hawaiian and Other Pacific Islander	462,534	0.16	398,835	0.14

Figure 12.7   Modified Race Data Summary table from the year 2000 United States Census.

Using Perl, Python, or Ruby, and the MR(31)-CO.txt file, write a script that calculates the populations for the same racial categories as shown in the summary table. Are your numbers close to the numbers provided in the table?

5. Using Perl, Python, or Ruby, modify the script you wrote for Exercise 3, listing the populations for the same racial categories but for each state.

13
CENTERS FOR DISEASE CONTROL AND PREVENTION MORTALITY FILES

The CDC (U.S. Centers for Disease Control and Prevention) prepares a public use data set containing the deidentified records of virtually every death occurring in the United States, over a period of a year. These yearly data files each exceed 1 GB in length, and contain several million records.

With access to the CDC mortality files, we can glean a wealth of information related to the immediate, underlying, and contributing causes of death in the United States.

Because each record contains multiple conditions related to the death of individuals, or conditions present in the individual at the time of death, it is possible to draw inferences about the relationships among the different conditions, and the likelihood of coexistences among conditions.

Because demographic information is provided in the mortality records, it is possible to determine the frequency of occurrence of conditions in age groups, ethnic groups, localities, and genders.

Because the records meticulously preserve the order and organization of the original death certificate, it is possible to relate conditions by their order of causation (which conditions lead to which other conditions).

Because the disease conditions are coded using International Classification of Diseases, Version 10 (ICD10), all of the disease entries can be understood and correlated with terms from any other data set, coded with the same nomenclature.

Because there are over 2.3 million records in the CDC data set, it is possible to find large numbers of cases for hundreds of different conditions.

Because every record conforms to a consistent organization, it is possible to reorganize and merge these records with data from other sources, increasing the value of the original data.

In this report, we will use simple, open source, and freely available techniques to analyze the public use CDC mortality files.

13.1 Death Certificate Data

Much of what we think we know about the ways that Americans die comes from analyses of death certificates. Annual death certificate data for the entire U.S. population have been collected since 1935 by the Vital Statistics Program of the National Center for Health Statistics. Death certificate data is notoriously error prone, and the problems seem to extend beyond national borders, as a similar set of complaints have been voiced in the United States and the United Kingdom. The most common error occurs when a mode of death is listed as the cause of death (e.g., cardiac arrest, cardiopulmonary arrest), thus nullifying the potential value of the death certificate. A recent survey of 49 national and international health atlases has shown that there is virtually no consistency in the way that death data is presented.

Members of the public may believe that death certificates are completed after a formal autopsy is conducted. This is seldom the case. Autopsies are conducted in only a small percentage of deaths worldwide. Autopsies can take weeks before the final report is issued. Doctors who complete the death certificate, usually within minutes or hours of the patient's death, do so without the benefit of a pathologist's postmortem examination. The death certificate contains a doctor's best guess of the patient's cause of death, but the best guess may be inaccurate.

Complicating the "cause of death" data is the rather strange ways we have come to think about the biological steps leading to death. For centuries, the cause of death has been encapsulated in a backwardly sequential list of conditions.

The example lists an underlying cause of death leading to an immediate cause of death:

a. Bleeding of esophageal varices
b. Portal hypertension
c. Liver cirrhosis
d. Hepatitis B

Hepatitis B is the underlying cause of death. Hepatitis led to the development of liver cirrhosis, which, in turn, produced portal hypertension. Portal hypertension led to the development of esophageal varices. The varices bled. The patient's proximate cause of death was internal bleeding (from esophageal varices). Hepatitis B was the antecedent for every condition listed.

How would this be entered on the patient's death certificate? Let us look at a blank form (Figure 13.1). This figure is extracted from a U.S. government publication available at

ftp.cdc.gov/pub/Health_Statistics/NCHS/Dataset_Documentation/mortality/
    Mort99doc.pdf

In Part 1 of Item 27, hepatitis B would be listed on line "d"; liver cirrhosis on line "c"; portal hypertension on line "b"; and bleeding of esophageal varices on line "a." Additional significant medical conditions that did not cause the patient's death are listed in Part 2 of Item 27. Nothing could be easier!

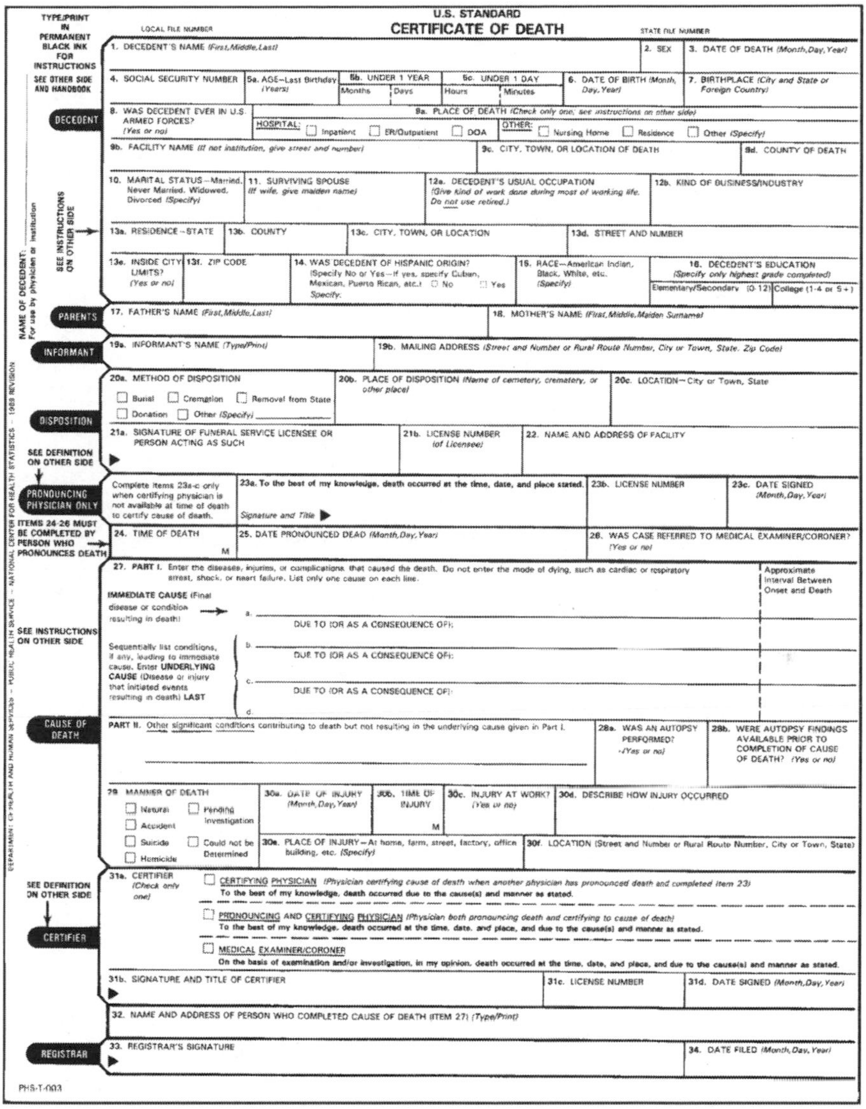

TYPE/PRINT IN PERMANENT BLACK INK FOR INSTRUCTIONS SEE OTHER SIDE AND HANDBOOK

LOCAL FILE NUMBER

U.S. STANDARD
CERTIFICATE OF DEATH

STATE FILE NUMBER

DECEDENT

1. DECEDENT'S NAME *(First, Middle, Last)*
2. SEX
3. DATE OF DEATH *(Month, Day, Year)*
4. SOCIAL SECURITY NUMBER
5a. AGE–Last Birthday *(Years)*
5b. UNDER 1 YEAR — Months | Days
5c. UNDER 1 DAY — Hours | Minutes
6. DATE OF BIRTH *(Month, Day, Year)*
7. BIRTHPLACE *(City and State or Foreign Country)*
8. WAS DECEDENT EVER IN U.S. ARMED FORCES? *(Yes or no)*
9a. PLACE OF DEATH *(Check only one; see instructions on other side)*
HOSPITAL: ☐ Inpatient ☐ ER/Outpatient ☐ DOA
OTHER: ☐ Nursing Home ☐ Residence ☐ Other *(Specify)*
9b. FACILITY NAME *(If not institution, give street and number)*
9c. CITY, TOWN, OR LOCATION OF DEATH
9d. COUNTY OF DEATH
10. MARITAL STATUS–Married, Never Married, Widowed, Divorced *(Specify)*
11. SURVIVING SPOUSE *(If wife, give maiden name)*
12a. DECEDENT'S USUAL OCCUPATION *(Give kind of work done during most of working life. Do not use retired.)*
12b. KIND OF BUSINESS/INDUSTRY
13a. RESIDENCE–STATE
13b. COUNTY
13c. CITY, TOWN, OR LOCATION
13d. STREET AND NUMBER
13e. INSIDE CITY LIMITS? *(Yes or no)*
13f. ZIP CODE
14. WAS DECEDENT OF HISPANIC ORIGIN? (Specify No or Yes–If yes, specify Cuban, Mexican, Puerto Rican, etc.) ☐ No ☐ Yes Specify:
15. RACE–American Indian, Black, White, etc. *(Specify)*
16. DECEDENT'S EDUCATION *(Specify only highest grade completed)* Elementary/Secondary (0-12) | College (1-4 or 5+)

NAME OF DECEDENT: For use by physician or institution

SEE INSTRUCTIONS ON OTHER SIDE

PARENTS

17. FATHER'S NAME *(First, Middle, Last)*
18. MOTHER'S NAME *(First, Middle, Maiden Surname)*

INFORMANT

19a. INFORMANT'S NAME *(Type/Print)*
19b. MAILING ADDRESS *(Street and Number or Rural Route Number, City or Town, State, Zip Code)*

DISPOSITION

20a. METHOD OF DISPOSITION ☐ Burial ☐ Cremation ☐ Removal from State ☐ Donation ☐ Other *(Specify)*
20b. PLACE OF DISPOSITION *(Name of cemetery, crematory, or other place)*
20c. LOCATION–City or Town, State
21a. SIGNATURE OF FUNERAL SERVICE LICENSEE OR PERSON ACTING AS SUCH ▶
21b. LICENSE NUMBER *(of Licensee)*
22. NAME AND ADDRESS OF FACILITY

SEE DEFINITION ON OTHER SIDE

PRONOUNCING PHYSICIAN ONLY

Complete items 23a-c only when certifying physician is not available at time of death to certify cause of death.
23a. To the best of my knowledge, death occurred at the time, date, and place stated. *Signature and Title* ▶
23b. LICENSE NUMBER
23c. DATE SIGNED *(Month, Day, Year)*

ITEMS 24-26 MUST BE COMPLETED BY PERSON WHO PRONOUNCES DEATH

24. TIME OF DEATH M
25. DATE PRONOUNCED DEAD *(Month, Day, Year)*
26. WAS CASE REFERRED TO MEDICAL EXAMINER/CORONER? *(Yes or no)*

SEE INSTRUCTIONS ON OTHER SIDE

CAUSE OF DEATH

27. PART I. Enter the diseases, injuries, or complications that caused the death. Do not enter the mode of dying, such as cardiac or respiratory arrest, shock, or heart failure. List only one cause on each line.
Approximate Interval Between Onset and Death
IMMEDIATE CAUSE (Final disease or condition resulting in death) → a.
DUE TO (OR AS A CONSEQUENCE OF):
Sequentially list conditions, if any, leading to immediate cause. Enter **UNDERLYING CAUSE** (Disease or injury that initiated events resulting in death) **LAST**
b.
DUE TO (OR AS A CONSEQUENCE OF):
c.
DUE TO (OR AS A CONSEQUENCE OF):
d.
PART II. Other significant conditions contributing to death but not resulting in the underlying cause given in Part I.
28a. WAS AN AUTOPSY PERFORMED? *(Yes or no)*
28b. WERE AUTOPSY FINDINGS AVAILABLE PRIOR TO COMPLETION OF CAUSE OF DEATH? *(Yes or no)*
29. MANNER OF DEATH ☐ Natural ☐ Pending Investigation ☐ Accident ☐ Suicide ☐ Could not be Determined ☐ Homicide
30a. DATE OF INJURY *(Month, Day, Year)*
30b. TIME OF INJURY M
30c. INJURY AT WORK? *(Yes or no)*
30d. DESCRIBE HOW INJURY OCCURRED
30e. PLACE OF INJURY–At home, farm, street, factory, office building, etc. *(Specify)*
30f. LOCATION (Street and Number or Rural Route Number, City or Town, State)

SEE DEFINITION ON OTHER SIDE

CERTIFIER

31a. CERTIFIER *(Check only one)*
☐ CERTIFYING PHYSICIAN *(Physician certifying cause of death when another physician has pronounced death and completed item 23)*
To the best of my knowledge, death occurred due to the cause(s) and manner as stated.
☐ PRONOUNCING AND CERTIFYING PHYSICIAN *(Physician both pronouncing death and certifying to cause of death)*
To the best of my knowledge, death occurred at the time, date, and place, and due to the cause(s) and manner as stated.
☐ MEDICAL EXAMINER/CORONER
On the basis of examination and/or investigation, in my opinion, death occurred at the time, date, and place, and due to the cause(s) and manner as stated.
31b. SIGNATURE AND TITLE OF CERTIFIER ▶
31c. LICENSE NUMBER
31d. DATE SIGNED *(Month, Day, Year)*
32. NAME AND ADDRESS OF PERSON WHO COMPLETED CAUSE OF DEATH (ITEM 27) *(Type/Print)*

REGISTRAR

33. REGISTRAR'S SIGNATURE ▶
34. DATE FILED *(Month, Day, Year)*

PHS-T-003

DEPARTMENT OF HEALTH AND HUMAN SERVICES – PUBLIC HEALTH SERVICE – NATIONAL CENTER FOR HEALTH STATISTICS – 1989 REVISION

Figure 13.1 Prototypical death certificate. Item 27 collects the causes of death, Part 1, and other significant conditions, Part 2.

Seemingly intractable problems arise when

There are multiple, sometimes unrelated, conditions that contribute to the patient's death.
The doctor filling out the death certificate is not familiar with the patient's history.
The doctor has not been trained in the proper procedure for completing the death certificate.
The doctor does not make the effort to provide a complete and accurate death certificate.
The cause of death is obscure or contentious.
The doctor has a reason to conceal conditions leading to the cause of death.

Thousands of instructional pages have been written on the proper way to complete a death certificate. Though we strive to do our best, it is unlikely that humans can be expected to prepare consistent and accurate summaries of what has always been a phenomenon shrouded by ignorance.

In the next section, we will discuss how the data on every death certificate is transformed into a mortality record consisting of an alphanumeric sequence.

13.2 Obtaining the CDC Data Files

The CDC data sets are available by anonymous ftp from the ftp.cdc.gov server. Most browsers come equipped for the ftp protocol, and you can just enter the ftp address much as you might enter an http protocol Web address.

The address for the mortality data site is

ftp.cdc.gov/pub/Health_Statistics/NCHS/Datasets/DVS/mortality/

The CDC server's subdirectory is shown in Figure 13.2.

For medical data miners, this is a very important server site. When these files are unzipped, they provide an aggregate database of deidentified records, collected over several decades, of information culled from many millions of death certificates. This

File	Size	Time
Mort1968us.zip	37401 KB	10:06:00 AM
Mort1969us.zip	38401 KB	10:08:00 AM
Mort1970us.zip	40379 KB	10:09:00 AM
Mort1971us.zip	40519 KB	10:10:00 AM
Mort1972us.zip	22510 KB	10:11:00 AM
Mort1973us.zip	44418 KB	10:12:00 AM
Mort1974us.zip	44386 KB	10:14:00 AM
Mort1975us.zip	44837 KB	10:16:00 AM
Mort1976us.zip	43194 KB	10:17:00 AM
Mort1977us.zip	43042 KB	10:18:00 AM
Mort1978us.zip	41955 KB	10:20:00 AM
Mort1979us.zip	57476 KB	11:02:00 AM
Mort1980us.zip	57156 KB	11:03:00 AM
Mort1981dus.zip	24699 KB	10:21:00 AM
Mort1981us.zip	49214 KB	11:04:00 AM
Mort1982dus.zip	26015 KB	10:22:00 AM
Mort1982us.zip	50251 KB	11:06:00 AM
Mort1983us.zip	63120 KB	11:07:00 AM

Figure 13.2   Ftp index of the CDC yearly mortality files.

site alone can keep an epidemiologist busy and productive for his or her entire career. There is no limit to the utility of this site, when its data is merged with data from other biomedical resources. Using these files is no different than making a cake from a recipe: you assemble your ingredients, follow a series of steps, wait a few moments for the cake to cook, and enjoy the results.

The key file that we will be using is available by anonymous ftp from the CDC server:

ftp.cdc.gov/pub/Health_Statistics/NCHS/Datasets/mortalityMort99us.zip
   (88,077,536 bytes)

This file unzips to: Mort99us.dat (1,058,532,982 bytes)

Each record has the following general structure:

   0 1101999363010199999991363010329911540l 10111073402009 6 1010075
      990999 99999 199901015010150450 009 7 J4492670008606228005ll J969
      12J449 61E109 62I709 63I500 03 C259 E149 I10

This example is a composite record selected from string sequences in several different records. It is not necessary or appropriate to show an actual record from the CDC file.

Looking at the record, a seeming jumble of alphanumerics and spaces, you might conclude that extracting any useful information would be a formidable task, well beyond the capacity of nonspecialists. Actually, all of the data in the 1 GB file can be parsed, reassembled, and analyzed in a matter of seconds, with a few lines of code that anyone can understand and implement.

The CDC provides a data dictionary file that explains the meaning of each byte location in each mortality record:

ftp.cdc.gov/pub/Health_Statistics/NCHS/Dataset_Documentation/mortality/
   Mort99doc.pdf

The diseases listed in the CDC records are encoded as ICD10 (International Classification of Diseases, version 10) alphanumerics. We will need to have a nomenclature file to translate the ICD10 codes to English terms.

We will need the compilation of ICD codes prepared in Chapter 6, Section 6.1.

13.3 How Death Certificates Are Represented in Data Records

In this section, we will discuss how the data on every death certificate is transformed into a mortality record consisting of an alphanumeric sequence.

We will use the large (1 GB) CDC public mortality file and its data dictionary.

Mort99us.dat (1,058,532,982 bytes)

and

Mort99doc.pdf (4,911,017 bytes)

We could have just as easily used mortality data from other years. They are also available from the CDC ftp site.

The Mort99us.dat file consists of millions of records, one record per line, each line consisting of a long string of alphanumerics.

The portion of the line record sequence that we are most interested in is the stretch of alphanumerics extending from bytes 162 to 301.

The data dictionary file, on page 36, explains the significance of this stretch of characters (Figure 13.3).

Each 7-digit piece of this stretch of characters represents another diagnosis and consists of:

First character. Line indicator: The first byte represents the line of the death certificate on which the code appears. Six lines (1–6) are allowable, with the fourth and fifth lines denoting that an additional condition was written in beyond the four lines provided in Part I of the U.S. Standard Certificate of Death. Line 6 represents Part II of the death certificate.

Second character. Position indicator: The next byte indicates the position of the code on the line; that is, it is the first (1), second (2), third (3), ..., eighth (8) code on the line.

Third through sixth character. These four bytes represent the ICD10 (International Classification of Disease, version 10) code.

Seventh character. The seventh and last byte is blank.

This protocol permits us to capture all of the information conveyed in the cause of death section from the death certificate, including line number and number of causes on the line. The highest numbered cause of death line number (5 is the highest permissible number) indicates the underlying cause of death that leads, ultimately, to the proximate cause of death.

An example of a cause of death record sequence is

11I219 21I251 61I500 62R54

In this example, there are two causes of death:

11I219 (first line, first condition on line, ICD diagnosis I219)

and

21I251 (second line, first condition on line, ICD diagnosis I251)

In addition, there are two medical conditions that the doctor listed as "other significant conditions" that were not listed with the underlying causes of death (these are always designated with a "6").

1999
Mortality Multiple Cause-of-Death Public Use Record

Tape Location	Field Size	Item and Code Outline
162–301	140	ENTITY - AXIS CONDITIONS

Space has been provided for maximum of 20 conditions. Each condition takes 7 positions in the record. The 7th position will be blank. Records that do not have 20 conditions are blank in the unused area.

Position 1: Part/line number on certificate

1	...	Part I, line 1 (a)
2	...	Part I, line 2 (b)
3	...	Part I, line 3 (c)
4	...	Part I, line 4 (d)
5	...	Part I, line 5 (e)
6	...	Part II

Position 2: Sequence of condition within part/line

1–7	...	Code range

Position 3–6: Condition code

See Table 1 for a complete list of codes

162–168	7	1st Condition
169–175	7	2nd Condition
176–182	7	3rd Condition
183–189	7	4th Condition
190–196	7	5th Condition
197–203	7	6th Condition
204–210	7	7th Condition
211–217	7	8th Condition
218–224	7	9th Condition
225–231	7	10th Condition
232–238	7	11th Condition
239–245	7	12th Condition

Figure 13.3 Data dictionary describing byte locations for diseases listed in the death certificate.

61I500 ("other significant condition" list, item one, ICD code I500)
62R54 ("other significant condition" list, item two, ICD code R54)

The file does not tell us the term equivalent of the listed codes.
For this, we need to use an ICD10 dictionary.

I219 = (I21.9, Acute myocardial infarction unspecified)
I251 = (I25.1 Atherosclerotic heart disease)
I500 = (I50.0 Congestive heart failure)
R54 = (R54 Senility)

Note that the actual ICD10 codes contained a dot, and the CDC mortality sequence did not.

So, now we see the full picture of the cause-of-death section of the death certificate. Atherosclerotic heart disease was considered the underlying cause of death. Acute myocardial infarction was considered the proximate cause of death. Congestive heart failure and senility were considered "other significant conditions."

13.4 Ranking, by Number of Occurrences, Every Condition in the CDC Mortality Files

When we have the collected death certificates for the U.S. population (i.e., the CDC mortality files), the byte locations for the causes of death in each record (as a list of ICD codes), and a dictionary that translates ICD codes into medical terms, we can easily collect a list of causes of death and their frequency.

13.4.1 Script Algorithm

1. Open the ICD nomenclature file (each10.txt, described in Chapter 6, Section 6.1), and load the entire file into a string object.
2. Purge the string object of non-ASCII characters.
3. Create a text array by splitting the file wherever a newline character is followed by an ICD code (an uppercase letter, followed by up to five digits, including a "." character).
4. For each item in the array, extract the code and the corresponding term.
5. Create a dictionary object with ICD codes as keys and corresponding terms as values.
6. Close the ICD nomenclature file (each10.txt), and open the 1 GB U.S. mortality file for 1999. This is the Mort99us.dat file, which I happen to keep in my hard drive's c:\big subdirectory.
7. Parse through the mortality file, line by line.
8. For each parsed line, extract bytes 162 to 302, containing the list of ICD-coded conditions listed for each death certificate record.

9. Each code in the list of conditions is followed by a space. Split the codes on the space, and place them into an array object.
10. Create a counting dictionary object, with each encountered ICD code as a key, and the number of occurrences of the code (in the mortality file) as the value.
11. Each time a code is encountered, increment its value by 1.
12. After the mortality file is parsed, collect the key–value pairs in the counting dictionary object as string items in an array.
13. Print the sorted array items, along with the medical term associated with each ICD code.

Perl Script

We will use this file in the next Perl script, to determine the total number of each condition appearing in the 1 gigabyte mort99us.dat file. You will need to place the Mort99us.dat file in the same subdirectory as this Perl script.

```
#!/usr/local/bin/perl
open (ICD, "c\:\\ftp\\each10.txt")||die"cannot";
undef($/);
$line = <ICD>;
$line =~ tr/\000-\011//d;
$line =~ tr/\013-\014//d;
$line =~ tr/\016-\037//d;
$line =~ tr/\041-\055//d;
$line =~ tr/\173-\377//d;
@linearray = split(/\n(?=[ ]*[A-Z][0-9\.]{1,5})/, $line);
foreach $thing (@linearray)
  {
  if ($thing =~ /^ *([A-Z][0-9\.]{1,5}) ?/)
    {
    $code = $1;
    $term = $';
    $term =~ s/\n//;
    $term =~ s/[ ]+$//;
    $code =~ s/\.//;
    $dictionary{$code} = $term;
    }
  }
close ICD;
$/ = "\n";
open (ICD, "c\:\\big\\Mort99us\.dat");
$line = " ";
while ($line ne "")
  {
  $line = <ICD>;
  $codesection = substr($line,161,140);
  $codesection =~ s/ *$//;
```

```perl
@codearray = split(/ +/,$codesection);
foreach $code (@codearray)
   {
   $code =~ /[A-Z][0-9]+/;
   $code = $&;
   $counter{$code}++;
   }
}
close ICD;
open (OUT, ">cdc.out");
while ((my $key, my $value) = each(%counter))
   {
   $value = "000000" . $value;
   $value = substr($value,-6,6);
   push(@filearray, "$value  $key    $dictionary{$key}");
   }
$outfile = join("\n", reverse(sort(@filearray)));
print OUT $outfile;
exit
```

Python Script

```python
#!/usr/local/bin/python
import re, string
icd_file = open("c:\\ftp\\each10.txt", "r")
icd_string = icd_file.read()
line_array = re.split(r'\n(?= *[A-Z][0-9\.]{1,5})', icd_string)
dictionary = {}
counter = {}
codearray = []
results_array = []
for thing in line_array:
   thing_match = re.search(r'^ *([A-Z][0-9\.]{1,5}) ?(.+)$', thing)
   if thing_match:
     code = thing_match.group(1)
     term = thing_match.group(2)
     term = re.sub(r'[^a-zA-Z ]',"", term)
     term = string.rstrip(term)
     code = re.sub(r'\.',"", code)
     dictionary[code] = term
mort_txt = open("c:\\big\\mort99us.dat", "r")
for line in mort_txt:
   codesection = line[161:302]
   codesection = re.sub(r' *$', "", codesection)
   codearray = re.split(r' +', codesection)
   for code in codearray:
     code_match = re.search(r'([A-Z][0-9]+)', code)
     if code_match:
        code = code_match.group(1)
```

```
      if dictionary.has_key(code):
        if counter.has_key(code):
          counter[code] = int(counter[code]) + 1
        else:
          counter[code] = 1
mort_txt.close()
out_mort = open("cdc.out", "w")
for key,value in counter.iteritems():
    value = str(value)
    value = "000000" + value
    value = value[-6:]
    results_array.append(value + " " + key + " " + dictionary[key])
results_array.sort()
results_array.reverse()
print>>out_mort, '\n'.join(results_array)
exit;
```

Ruby Script

```ruby
#!/usr/local/bin/ruby
icd_string = IO.read("c:/ftp/each10.txt")
linearray = icd_string.split(/\n(?= *[A-Z][0-9\.]{1,5})/)
dictionary = {}
counter = Hash.new(0)
codearray = []
results_array = []
linearray.each do
  |thing|
  if (thing =~ /^ *([A-Z][0-9\.]{1,5}) ?/)
    code = $1
    term = $'
    term.sub!(/[^a-zA-Z ]/,"")
    term.sub!(/[ ]+$/,"")
    code.sub!(/\./,"")
    dictionary[code] = term
  end
end
mort_txt = File.open("c:/big/Mort99us.dat", "r")
mort_txt.each_line do
  |line|
  codesection = line.slice(161,140)
  codesection.sub!(/ *$/, "")
  codearray = codesection.split(/ +/)
  codearray.each do
    |code|
    code =~ /[A-Z][0-9]+/
    code = $&
    if dictionary.has_key?(code)
      counter[code] = counter[code] + 1
```

```
      end
    end
end
mort_txt.close
out_mort = File.open("cdc.out", "w")
counter.each_pair do
    |key, value|
    value = value.to_s
    value = "000000" + value
    value = value.slice(-6,6)
    results_array.push(value + " " + key + " " + dictionary[key])
end
out_mort.puts results_array.sort.reverse.join("\n")
exit
```

13.4.2 Analysis

On my 2.5 GHz CPU/512 megabyte (MB) RAM desktop computer, it takes well under a minute to parse through the 1 gigabyte (GB) CDC mortality data set and produce the desired output file (cdc.out). The total number of records parsed by the script was 2,394,871. There are 5,650 different conditions included in the 1999 CDC mortality data set.

The first lines of the output file are shown in Figure 13.4.

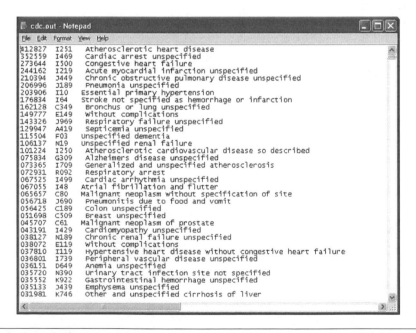

Figure 13.4   A sample of script output, listing the number of occurrences of a disease in the CDC mortality records (in descending order), followed by the ICD code for the diagnosis, followed by the term name for the diagnosis.

The top line is

412827 I251 Atherosclerotic heart disease

It indicates that atherosclerotic heart disease is the most common condition listed in the death certificates in 1999 in the United States. It was listed 412,827 times. The ICD10 code for atherosclerotic heart disease is I25.1.

Some of the output lines do not seem particularly helpful. For example:

000456 D487 Other specified sites
000451 C482 Peritoneum unspecified
000448 C210 Anus unspecified

Nobody dies from "Peritoneum unspecified." The strange diagnosis is explained by the rather unsatisfactory way that the ICD assigns terms to codes. In this case, Peritoneum unspecified is a subterm in the general category of "C48, Malignant neoplasm of retroperitoneum and peritoneum."

Whenever an ICD term appears uninformative, we can return to the each10.txt file and clarify its meaning by examining the root term for the subterm.

Exercises

1. Rank by the number of occurrences of death certificate diseases by gender.
2. Rank by the number of occurrences of death certificate diseases by ethnicity.
3. Rank by the number of occurrences of death certificate diseases by ages (divided into 10-year intervals).
4. Determine the disease that is listed most often as a co-occurring condition in the list of diseases contained in U.S. death certificates. That is, if a death certificate lists more than one disease, what disease is most likely to be included among the multiple-occurring diseases? Is this the same disease as the most commonly occurring disease listed on U.S. death certificates?

PART III
PRIMARY TASKS OF MEDICAL INFORMATICS

14
AUTOCODING

In the field of biomedical informatics, it is often necessary to extract medical terms from text and attach a nomenclature concept code to the extracted term. By doing so, concepts of interest contained in text can be retrieved regardless of the choice of words used to describe them. For example, hepatocellular carcinoma, liver cell cancer, liver cancer, and hcc might all be given the same code number in a neoplasm nomenclature. Documents using any of these terms can be collected and merged if all of the terms are annotated with the same concept code. A software product that computationally parses and codes medical text is called an *autocoder* or an *automatic coder*.

Many people believe that it is difficult to write autocoding software (that can parse text, find and extract medical terms, and add nomenclature codes to terms). Furthermore, many people believe that it is impossible to write fast autocoding software. People accept autocoder speeds that code a typical pathology report at a rate of one report (about 1,000 bytes) per second

Both of these notions are false. A superb autocoder can be written in a few dozen lines of code. In this chapter, we will write a short, simple autocoding script that improves on the rate of other autocoders by at least 100-fold.

14.1 A Neoplasm Autocoder

The script requires two external files, neocl.xml, the Neoplasm Classification in XML format, available for download as a gzipped file from

http://www.julesberman.info/neoclxml.gz

There are about 135,000 unique terms in the nomenclature. Each term is listed in a consistent format, as shown in these two examples:

<name nci-code = "C3084300">polymorphous haemangioendothelioma</name>
<name nci-code = "C3085000">angioma</name>

The terms are enclosed by angle brackets:

>polymorphous haemangioendothelioma<
>angioma<

The codes are enclosed by quotations:

"C3084300"
"C3085000"

Terms and corresponding codes can be easily extracted by a simple regex expression.

We will use an external file that we can autocode. For this sample project, we will parse through tumorabs.txt, a file of 20,000 abstract titles extracted from PubMed and available for download at

http://www.julesberman.info/book/tumorabs.txt

A portion of the file is shown in Figure 14.1.

We described the process of obtaining PubMed search result files in Chapter 9, Section 9.1.

14.1.1 Script Algorithm

1. Open the nomenclature file, which will be the source of coded terms to match against the text that needs to be autocoded. For this example, we will use the neoplasm taxonomy, but it could be any nomenclature that consists of codes listed with their corresponding medical terms.
2. Create a dictionary object with keys corresponding to the terms (names of neoplasms, in this case) of the medical nomenclature and values comprising the corresponding codes for the terms.
3. Open the file to be parsed (tumorabs.txt).
4. Parse through the file, line by line, each line containing a sentence.
5. As each sentence is parsed, break the file into every possible ordered subsequence of words (a phrase array). For example, "Everybody loves to eat pizza" would be broken into an array containing the following items:
   Everybody loves to eat pizza
   Everybody loves to eat
   Everybody loves
   Everybody
   loves to eat pizza
   loves to eat
   loves to
   loves
   to eat pizza
   to eat
   to
   eat pizza
   eat
   pizza

```
local versus diffuse recurrences of meningiomas factors correlated to the extent of the recurrence
the effect of an unplanned excision of a soft tissue sarcoma on prognosis
obstructive jaundice associated burkitt lymphoma mimicking pancreatic carcinoma
efficacy of zoledronate in treating persisting isolated tumor cells in bone marrow in patients with breast cance
metastatic lymph node number in epithelial ovarian carcinoma does it have any clinical significance
extended three dimensional impedance map methods for identifying ultrasonic scattering sites
aberrant expression of connexin 26 is associated with lung metastasis of colorectal cancer
microrna expression profiles of esophageal cancer
state and trait anxiety and depression in patient with primary brain tumors before and after surgery 1 year lon
laparoscopic resection of large adrenal ganglioneuroma
case records of the massachusetts general hospital case 4 2008 a 33 year old pregnant woman with swelling of the
evaluation of higher order time domain perturbation theory of photon diffusion on breast equivalent phantoms and
meningeal melanocytosis in a young patient an autopsy diagnosis
oncogenic hypophosphataemic osteomalacia biomarker roles of fibroblast growth factor 23 1 25 dihydroxyvitamin d3
microrna expression profiles associated with prognosis and therapeutic outcome in colon adenocarcinoma
manifestation of malakoplakia in a urethral diverticulum in a female patient
six versus eight cycles of bi weekly chop 14 with or without rituximab in elderly patients with aggressive cd20
giant abdominal tumor of the ovary
an up to date anti cancer treatment strategy focusing on hif 1alpha suppresion its application for refractory
obesity alters cytokine gene expression and promotes liver injury in rats with acute pancreatitis
intra cardiac lymphoma with right heart failure a therapeutic emergency in two patients
```

Figure 14.1 The first few lines of the tumorabs.txt file, with each sentence assigned to a separate line of the file.

6. For each item in the phrase array, determine whether the item matches a term in the neoplasm dictionary object.

7. If there is a match, print the phrase and the corresponding code to an external file.

8. The external file will consist of the lines from the text, followed by the phrases from the lines that are neoplasm terms, along with their nomenclature codes.

Perl Script

```perl
#!/usr/local/bin/perl
open(TEXT, "neocl.xml");
$line = " ";
while ($line ne "")
    {
    $line = <TEXT>;
    next if ($line !~ /\"(C[0-9]{7})\"/);
    $line =~ /\"(C[0-9]{7})\"/;
    $code = $1;
    $line =~ /\"\> ?(.+) ?\<\//;
    $phrase = $1;
    if ($phrase =~ /[a-z]/)
        {
        $literalhash{$phrase} = $code;
        }
    }
$phrase = "";
close TEXT;
print "Neoplasm code hash has been created.  Autocoding will start
now\n";
open(ABSFILE, "tumorabs.txt")||die"cannot";
open(OUTFILE, ">tumorab2.out")||die"cannot";
$line = " ";
while($line ne "")
    {
    $line = <ABSFILE>;
    $sentence = $line;
    $sentence =~ s/\n//o;
    $sentence =~ s/omas/oma/g;
    $sentence =~ s/tumo[u]?rs/tumor/g;
    print OUTFILE "\nTitle..." . ucfirst($sentence) . "." . "\n";
    @sentence_array = split(/ /,$sentence);
    $cycles = scalar(@sentence_array);
    for($n=0;$n<$cycles;$n++)
        {
        for($i=0;$i<scalar(@sentence_array);$i++)
            {
            @part_array = @sentence_array[0..$i];
            $phrase = join(" ", @part_array);
            if (exists($literalhash{$phrase}))
```

```
            {
            print OUTFILE "Autocoded tumor..." . ucfirst($phrase)
. " " . $literalhash{$phrase} . "\n";
            }
          }
        shift(@sentence_array);
        }
      }
   }
exit;
```

Python Script

```python
#!/usr/local/bin/python
import re
text = open("neocl.xml", "r")
literalhash = {}
codematch = re.compile('\"(C\d{7})\"')
phrasematch = re.compile('\"\> ?(.+) ?\<\/')
for line in text:
    m= codematch.search(line)
    if m:
      code = m.group(1)
    else:
      continue
    x = phrasematch.search(line)
    if x:
      phrase = x.group(1)
    else:
      continue
    literalhash[phrase] = code
text.close()
print "Neoplasm code hash has been created.  Autocoding will start
now"
absfile = open("tumorabs.txt", "r")
outfile = open("tumorpy.out", "w")
singular = re.compile('omas')
england = re.compile('tumo[u]?rs')
for line in absfile:
    sentence = line
    sentence = singular.sub("oma",sentence)
    sentence = england.sub("tumor",sentence)
    sentence = sentence.rstrip()
    print>>outfile,"\nAbstract title..." + sentence + "."
    sentence_array = sentence.split(" ")
    length = len(sentence_array)
    for i in range(length):
      for place_length in range(len(sentence_array)):
          last_element = place_length + 1
          phrase = ' '.join(sentence_array[0:last_element])
```

```
        if literalhash.has_key(phrase):
             print>>outfile,"Neoplasm term..." + phrase + " " +
literalhash[phrase]
        sentence_array.pop(0)
exit
```

Ruby Script

```ruby
#!/usr/local/bin/ruby
text = File.open("neocl.xml", "r")
literalhash = Hash.new
text.each do
    |line|
    next if (line !~ /\"(C[0-9]{7})\"/)
    line =~ /\"(C[0-9]{7})\"/
    code = $1;
    line =~ /\"\> ?(.+) ?\<\//
    phrase = $1;
    if (phrase =~ /[a-z]/)
        literalhash[phrase] = code
        #puts phrase
    end
end
text.close
puts "Neoplasm code hash has been created.  Autocoding will start
now"
absfile = File.open("tumorabs.txt", "r")
outfile = File.open("tumorabs.out", "w")
absfile.each do
    |sentence|
    sentence.chomp!
    sentence.gsub!(/omas/, "oma")
    sentence.gsub!(/tumo[u]?rs/, "tumor")
    outfile.puts "\nAbstract title..." + sentence.capitalize + "."
    sentence_array = sentence.split
    length = sentence_array.size
    length.times do
        (1..sentence_array.size).each do
            |place_length|
            phrase = sentence_array.slice(0,place_length).join(" ")
            if literalhash.has_key?(phrase)
                outfile.puts "Neoplasm term..." + phrase.capitalize +
" " + literalhash[phrase]
            end
        end
    sentence_array.shift
    end
end
exit
```

14.1.2 Analysis

The output of the coder is virtually perfect. Browse through the 10,000 abstract titles on this page and look for the named neoplasms in the abstract text. See if you can find named neoplasms included in the abstract title that were excluded from the autocoded terms that follow each abstract title.

Each abstract line parsed from the tumorabs.txt file is printed and then followed by the list of autocoded terms extracted from the title.

Note that the terms coded "C0000000" are general neoplasm terms such as "tumor" or "cancer" and not specific names of neoplasms, or they are names of neoplasms that have not yet been classified within the neoplasm taxonomy. Also, the program codes each occurrence of a neoplasm term, even if it is repeated.

> Abstract title. Local versus diffuse recurrences of meningioma factors correlated to the extent of the recurrence.
> Neoplasm term. Meningioma C3230000.

> Abstract title. The effect of an unplanned excision of a soft tissue sarcoma on prognosis.
> Neoplasm term. Soft tissue sarcoma C9306000.
> Neoplasm term. Sarcoma C0000000.

> Abstract title. Obstructive jaundice associated burkitt lymphoma mimicking pancreatic carcinoma.
> Neoplasm term. Jaundice C0000000.
> Neoplasm term. Burkitt lymphoma C7188000.
> Neoplasm term. Lymphoma C7065000.
> Neoplasm term. Pancreatic carcinoma C3850000.
> Neoplasm term. Carcinoma C0000000.

> Abstract title. Efficacy of zoledronate in treating persisting isolated tumor cells in bone marrow in patients with breast cancer a phase II pilot study.
> Neoplasm term. Tumor C0000000.
> Neoplasm term. Breast cancer C4872000.
> Neoplasm term. Cancer C0000000.

> Abstract title. Metastatic lymph node number in epithelial ovarian carcinoma does it have any clinical significance.
> Neoplasm term. Epithelial ovarian carcinoma C4908000.
> Neoplasm term. Ovarian carcinoma C4908000.
> Neoplasm term. Carcinoma C0000000.

Abstract title. Extended three-dimensional impedance map methods for identifying ultrasonic scattering sites.

Abstract title. Aberrant expression of connexin 26 is associated with lung metastasis of colorectal cancer.
Neoplasm term. Colorectal cancer C5105000.
Neoplasm term. Cancer C0000000.

Abstract title. Microrna expression profiles of esophageal cancer.
Neoplasm term. Esophageal cancer C3513000.
Neoplasm term. Cancer C0000000.

Abstract title. State and trait anxiety and depression in patients with primary brain tumor before and after surgery 1 year longitudinal study.
Neoplasm term. Primary brain tumor C0000000.
Neoplasm term. Brain tumor C0000000.
Neoplasm term. Tumor C0000000.

Abstract title. Laparoscopic resection of large adrenal ganglioneuroma.
Neoplasm term. Ganglioneuroma C3049000.

Abstract title. Case records of the Massachusetts general hospital case 4 2008 a 33- year-old pregnant woman with swelling of the left breast and shortness of breath.

Abstract title. Evaluation of higher order time domain perturbation theory of photon diffusion on breast equivalent phantoms and optical mammograms.

Abstract title. Meningeal melanocytosis in a young patient an autopsy diagnosis.

Abstract title. Oncogenic hypophosphataemic osteomalacia biomarker roles of fibroblast growth factor 23 1 25 dihydroxyvitamin d3 and lymphatic vessel endothelial hyaluronan receptor 1.

Abstract title. Microrna expression profiles associated with prognosis and therapeutic outcome in colon adenocarcinoma.
Neoplasm term. Colon adenocarcinoma C4349000.
Neoplasm term. Adenocarcinoma C0000000.

14.2 Recoding

> I may be the world's worst writer, but I'm the world's best rewriter.
>
> **—James Michener**

The medical informatics literature has lots of descriptions of medical autocoders, but most of these descriptions fail to include their speed. The autocoder included here is fast, coding 20,000 citations in about 20 seconds or less on my 2.5 GHz desktop CPU with 512 megabytes [MB] RAM). This is a rate of about 100 kilobytes per second. By the time this book is published, most readers will have computers that operate much faster than mine, providing a much faster autocoding rate.

Why is it important to have a fast autocoder? Why can't you load your parser with a big file and let it run in the background, taking as long as it takes to finish?

There are three reasons why you absolutely must have a fast autocoder:

1. Medical files today are large. It is not unusual for a large medical center to generate a terabyte of data each week. A slow autocoder could never keep up with the volume of medical information that is produced each day.
2. Autocoders, and the nomenclatures they draw terms from, need to be modified to accommodate unexpected oddities in the text that they parse (particularly formatting oddities and the inclusion of idiosyncratic language to express medical terms). The cycles of running a program, reviewing output, making modifications in software or nomenclatures, and repeating the whole process many times cannot be undertaken if you need to wait a week for your autocoding software to parse your text.
3. Autocoding is as much about recoding as it is about the initial process of providing nomenclature codes

You need to recode (supply a new set of nomenclature codes for terms in your medical text) whenever you want to change from one nomenclature to another.

You need to recode whenever you introduce a new version of a nomenclature.

You need to recode whenever you want to use a new coding algorithm (e.g., parsimonious coding versus comprehensive, or linking code to a particular extracted portion of report).

You need to recode whenever you add legacy data to your laboratory information systems.

You need to recode whenever you merge different medical data sets (especially, medical data sets that have been coded with different medical nomenclatures).

All of this recoding adds to the data burden placed on a medical autocoder.

It has been my personal observation that computational tasks that take much time (more than a few seconds) tend to be put on the back burner. The same observations would apply to medical deidentification software (Chapter 15), software designed to classify data into related groups (so-called intelligent computing) and software that draws inferences from classes of data (so-called artificial intelligence). Smart informaticians understand that program execution speed is always very important.

Exercises

1. Using Perl, Python, or Ruby, write a script that inserts the code, in parenthesis, immediately following each encountered neoplasm term in the file cancer_gene_titles.txt (created in Chapter 9, Section 9.1, and available at

   http://www.julesberman.info/book/cancer_gene_titles.txt

   Use the ICD-Oncology nomenclature, available at

   http://www.julesberman.info/book/icdo3.txt

   Additional information on the ICD-Oncology nomenclature is available in the appendix.
2. Repeat Exercise 1, using the Neoplasm Classification file, available in gzipped form at

   www.julesberman.info/neoclxml.gz

   Additional file information is available in the appendix.
3. Using Perl, Python, or Ruby, write a script that collects the neoplasm terms that are present in ICD-O (i.e., the icdo3.txt file) and absent from the Neoplasm Classification (i.e., the neocl.xml file).
4. Do the opposite of Exercise 3. Using Perl, Python, or Ruby, write a script that collects the neoplasm terms that are present in the Neoplasm Classification (i.e., the neocl.xml file) and absent from ICD-O (i.e., the icdo3.txt file).
5. Using Perl, Python, or Ruby, write a script that collects all the neoplasm terms that are present in the Neoplasm Classification that have the word *precancer* in the term.
6. Modify the script from Exercise 5 to collect all the neoplasm terms that are present in the Neoplasm Classification that have the word *precancer* in the term, along with all of the terms that have the same code as any of the terms that contain the word *precancer* in the term. Remember, multiple synonymous or near-synonymous terms will have the same concept code.

15
TEXT SCRUBBER FOR DEIDENTIFYING CONFIDENTIAL TEXT

Throughout history, people have tried very hard to remove confidential, private, offensive, or otherwise objectionable text from documents. With chisel, stylus, pen, white-out, magic marker, or mouse in hand, legions of censors have been reading our most intimate letters and stories, eager to blot out expletives, formulas, locations, names, and times, in the hope that their efforts will render text safe to share.

Human censors do an adequate job when the data flow is small, but the amount of sensitive information created in our electronic age is immense. Large hospitals create terabytes of information every week, and a good portion of that information comes in the form of free text (i.e., unstructured text, or text not constrained to fields in a form or template). The medical records of patients are confidential. Those who want to use this information for research purposes have two options: (1) obtain informed consent from patients to use their records (an impossible task if you want to analyze data from thousands of human subjects), or (2) deidentify the records by removing any information that could link the contents of a medical record to an individual patient.

In the past several decades, a variety of programs have been written that attempt to automatically remove identifying, private, or objectionable information from medical records. These programs are sometimes called "scrubbers", and most of these programs use the following algorithm:

1. Prepare lists of patient names, hospital staff names, addresses, obscenities, objectionable hospital slang, and hospital identifier numbers.
2. Parse through the text, deleting or replacing entries from the list with non-informational characters.
3. Match the text against a series of regex patterns that might indicate the presence of identifying information (e.g., formalisms such as Mr., Dr., Mrs. followed by another word, or numeric values, or date components), and remove these strings.

These methods are the software equivalent of the human who reads through letters and documents and marks over the objectionable parts. Parsing scripts that pass documents through a long series of regex filters are always slow, and they never completely remove objectionable material. They merely reduce the occurrences of objectionable text, without eliminating the problem.

There is a better way that is essentially the reverse of censorship. You create a list of acceptable phrases, and you parse through the text, deleting everything that is not included on your list. This method can parse text very quickly, because it has no regex filters. The method is potentially perfect, because the only text that appears in the final document is text composed of words and phrases that were preapproved.

15.1 Script Algorithm

1. In Chapter 9, Section 9.2, we created a list of word doublets from a PubMed corpus, consisting of titles of research papers written on the subject of cancer genes. For this chapter, we created a similar doublet list, available for download at http://www.julesberman.info/book/doublets.txt.
2. Begin your script by prompting the user to enter a sentence. The user may feel free to enter a sentence that is offensive, incriminating, filled with the names of people, or with sensitive information.
3. The entered text is parsed, word doublet by word doublet, with each doublet consisting of every word in the text followed by the next consecutive word.
4. Comparisons are made against the list of preapproved doublets (doublets.txt in this case).
5. Word doublets in the text that match word doublets on the list are saved. Everything else is replaced by an asterisk.

Perl Script

```perl
#!/usr/local/bin/perl
open (TEXT, "c:\\ftp\\doublets.txt")||die"cannot";
$line = " ";
while ($line ne "")
   {
   $line = <TEXT>;
   $line =~ s/\n//o;
   $line =~ s/ +$//o;
   $doubhash{$line} = "";
   }
close TEXT;
print "What text would you like scrubbed?\n";
$line = <STDIN>;
$line =~ s/\n//;
$line = lc($line);
$phrase =~ s/\'s//g;
$phrase =~ s/\,/ /g;
$line =~ s/[^a-z0-9 \-]/ /g;
@hoparray = split(/ +/,$line);
for ($i=0;$i<(scalar(@hoparray));$i++)
   {
   $doublet = "$hoparray[$i] $hoparray[$i+1]";
```

```
    if (exists $doubhash{$doublet})
        {
        print " $hoparray[$i]";
        $lastword = " $hoparray[$i+1]";
        }
    else
        {
        print $lastword;
        $lastword = " \*";
        }
    }
exit;
```

Python Script

```
#!/usr/local/bin/python
import sys, re, string
doub_file = open("c:\\ftp\\doublets.txt", "r")
doub_hash = {}
for line in doub_file:
  line = string.rstrip(line)
  doub_hash[line] = " "
doub_file.close()
print "What would you like to scrub?"
line = sys.stdin.readline()
line = string.lower(line)
line = string.rstrip(line)
linearray = re.split(r' +', line)
lastword = "*"
for i in range(0, len(linearray)):
    doublet = " ".join(linearray[i:i+2])
    if doub_hash.has_key(doublet):
      print " " + linearray[i],
      lastword = " " + linearray[i+1]
    else:
      print lastword,
      lastword = " *"
    if (i == len(linearray) + 1):
      print lastword
exit
```

Ruby Script

```
#!/usr/local/bin/ruby
doub_file = File.open("c:/ftp/doublets.txt", "r")
doub_hash = {}
doub_file.each_line{|line| line.chomp!; doub_hash[line] = " "}
doub_file.close
puts "What would you like to scrub?"
linearray = gets.chomp.downcase.split
arraysize = linearray.length - 2
```

```
lastword = "*"
for arrayindex in (0 .. arraysize)
   doublet = linearray[arrayindex] + " " + linearray[arrayindex+1]
   if doub_hash.key?(doublet)
     print " " + linearray[arrayindex]
     lastword = " " + linearray[arrayindex+1]
   else
     print lastword
     lastword = " *"
   end
   if arrayindex == arraysize
     print lastword
   end
end
exit
```

15.2 Analysis

Sample output:

> Input: Dr. Frankenstein killed his patient.
> Output: * * * *
>
> Input: The patient refused treatment
> Output: the patient * *
>
> Input: The patient has a poorly differentiated prostate carcinoma
> Output: the patient has a poorly differentiated prostate carcinoma
>
> Input: Sloan Kettering Hospital has admitted several patients with hepatomas.
> Output: * hospital has * several patients with *
>
> Input: Cancer can often be treated with surgery
> Output: cancer can often be treated with surgery

The doublet method script, with minor modifications, can scrub any length of any text. To illustrate, I downloaded a public domain book from Project Gutenberg.

Project Gutenberg is a remarkable resource that publishes plain-text versions of literary gems that have passed out of copyright. I used *Anomalies and Curiosities of Medicine* by George M. Gould and Walter Lytle Pyle. This book has lots of medical terminology and vaguely resembles the kind of text that might be included in a pathology report. Anyone can download the same text from

http://www.gutenberg.org/etext/747

An example of output paragraph is shown below. As expected with the doublet method, there are many blocked words. This is a limitation of the doublet method. If you use the standard list of doublets on any random book, you are bound to block some innocent doublets that were not included in the "approved" list. The only way to get around this limitation is to try to add safe doublets (from the text) to the approved list.

> In this important *, *, * * some historical *, describes a long series of experiments performed on * in order to * the passage of *, *, *, *, *, *, * * the placenta. The placenta shows a real affinity for * substances; in it * copper and mercury, but *, and it is therefore * it that the * * *; in addition to its *, intestinal, and *, * * glycogen and acts as an * *, and so resembles in its action the liver; * * of the fetus * only a potential *. * up of * in the placenta is not so general as * of them in the liver of the mother. It may be * the placenta does not form a barrier to the passage of * the circulation of the fetus; this would seem to * * *, which was always found in the * never in the fetal organs. In * * lead and * accumulation of the * in the fetal tissues is * in the maternal, perhaps from differences in * * or from greater diffusion. * it is * * barrier to the passage of *, * * * * degree of obstruction: it allows copper and * * *, * with greater difficulty. The * toxic substances in the fetus does not follow the same * * the adult. They * more widely in the fetus. In the * liver is the chief * *. *, which in * * to accumulate in the liver, is in the fetus * in the skin; copper accumulates in the fetal liver, * system, and sometimes in the skin; * which is * in the maternal liver, but also in the skin, has * in the skin, liver, * centers, and elsewhere * *. The frequent presence of * in the fetal * its physiologic importance. It has probably not * * influence on its *. On the * in the placenta and nerve * * * * abortion and the birth of dead *) Copper and lead did not cause *, * * so in two out of six *. Arsenic is a * agent in the *, * * * * *. An important * is that * * is frequently and seriously affected in syphilis, * * the special * for the accumulation of *. * * * * * action in this disease? The * of lead in the central nervous system of the * the frequency and serious character of * lesions. The presence of * in the * * * an explanation of the therapeutic results of * of this substance in skin *.

The deidentified output for the entire book is available at

http://www.julesberman.info/aacom10.htm

The strengths of the doublet method are accuracy and speed (the 2.4 megabyte [MB] book was deidentified in 3 seconds). I have never encountered an identifier (name of person, personal information, or any other data that can be linked to a specific person) in text scrubbed by the doublet method.

Exercises

1. In Section 15.1, we examined the output of the scrubber operating on a public domain book. We did not provide the script that does the job. Using Perl, Python, or Ruby, modify the script described in Section 15.1 to accept a book, or any plain-text file, as input.
2. Choose a book or plain-text file. You can use the same Project Gutenberg book used in Section 15.1, if you wish. Scrub the entire book, using the script you wrote in Exercise 1. Read the first few pages of the output. Can

you find any identifying terms (names of people, places, identifying code, etc.) in the output?

3. In Chapter 14, we described an autocoder. In this chapter, we described a scrubber. Why not do both at once?

   I have prepared a large text file of 95,260 citation titles. Each line of the file is an unpunctuated title. The titles all relate, in one way or another, to tumors, making the file suitable for autocoding with the Neoplasm Classification. The gzipped file, which you can use in this exercise, can be downloaded at http://julesberman.info/book/tumor_ti.gz.

   A script that autocodes and scrubs all 95,260 titles, in about 1 minute, has been prepared. It produced an output that consists of each original title, followed by the scrubbed version of the title, followed by the autocoded terms within the title. The text of the output of the combined autocoder/scrubber can be downloaded at http://julesberman.info/tu_both.gz.

   Some of the output can be viewed at http://www.julesbermsn.info/tu_both.htm (Figure 15.1).

   Using Perl, Python, or Ruby, write your own script that autocodes and scrubs text.

4. In Exercise 3, the scrubber and autocoder does not need to preserve the original punctuation of the reference title, because all of the titles in the sample text have been converted to lowercase, and have had their punctuation removed.

   A text file, with its original case and punctuation intact, is available at http://www.julesberman.info/book/pathol5.txt.

   Using Perl, Python, or Ruby, modify the combined autocoder and scrubber to produce an automatic autocoder and scrubber that uses an unmodified input text (with uppercase letters and punctuation) and preserves the original case and punctuation in the output text.

```
Original    -    carcinoid tumor of the common bile duct a rare complication of von hippel lindau syndrome
Scrubbed    -    carcinoid tumor of the common bile duct a rare complication of von * * *.

Autocode    -    carcinoid C4139100
Autocode    -    tumor C0000000
Autocode    -    carcinoid tumor C4139100

Original    -    establishment and characterization of a new cell line derived from human colorectal laterally
Scrubbed    -    * and characterization of a new cell line derived from human colorectal * * *.

Autocode    -    tumor C0000000

Original    -    in vivo anti tumor effect of hybrid vaccine of dendritic cells and esophageal carcinoma cells
Scrubbed    -    in vivo * * effect of * * * dendritic cells and esophageal carcinoma cells on esophageal carc

Autocode    -    tumor C0000000
Autocode    -    esophageal carcinoma C3513000
Autocode    -    carcinoma C0000000

Original    -    modeling intra tumor protein expression heterogeneity in tissue microarray experiments
Scrubbed    -    * * tumor protein expression heterogeneity in tissue * *.

Autocode    -    tumor C0000000

Original    -    caveolin 1 a tumor promoting role in human cancer
Scrubbed    -    * * a tumor * role in human cancer.

Autocode    -    cancer C0000000
Autocode    -    tumor C0000000
```

Figure 15.1 Web page showing the combined scrubbed and autocoded output for a list of reference titles.

16

WEB PAGES AND CGI SCRIPTS

There are many network protocols for exchanging information over the Internet, and for using remotely located applications. The number of standard protocols increases every day. The bad news is that if you are a healthcare worker, with limited programming skills and limited time for computer-related activities, you cannot master the field of distributed network computation.

Nonetheless, you should definitely learn the fundamentals of HTTP (HyperText Transfer Protocol) and CGI (Common Gateway Interface) programming. HTTP is the language that your browser uses to fetch Web pages from servers. Automating HTTP requests is extremely simple, and we provide a sample script in this chapter. Also, we will show you how CGI uses your own server-side scripts to grab information passed by a Web client (i.e., a browser), and return a Web page that is created on the fly, by your script, using any and all of the computational facilities available on the server. Once you have learned how to build a CGI script, all of the interactive, informational, and computational potential of the Internet is at your service.

16.1 Grabbing Web Pages

Accessible Web pages are files (usually in HTML format) that reside on servers which accept HTTP requests from clients connected to the Internet. Browsers are software applications that send HTTP requests and display the received Web pages. Using Perl, Python, or Ruby, you can automate HTTP requests. For each language, the easiest way to make an HTTP request is to use a module that comes bundled as a standard component of the language.

16.1.1 Script Algorithm

1. Import the module that makes HTTP requests.
2. Make the HTTP request.
3. If the request returns the Web page, print the page. Otherwise, print a message indicating the request was unsuccessful.

Perl Script

```
#!/usr/local/bin/perl
use LWP::Simple;
$good_url = qq|http://julesberman.info/factoids/batch.htm|;
```

```perl
$content = get($good_url);
if (defined ($content))
  {
  print $content;
  }
else
  {
  print "\nSorry, the get() call returned undef for $good_url";
  }
$bad_url = qq|http://julesberman.info/factoids/xxxxx.htm|;
$content = get($bad_url);
if (defined ($content))
  {
  print $content;
  }
else
  {
  print "\nSorry, the get() call returned undef for $bad_url";
  }
exit;
```

For Perl, the module is LWP::Simple. A Web page that explains the module syntax is available at

   http://search.cpan.org/~gaas/libwww-perl-5.834/lib/LWP/Simple.pm

Python Script

```python
#!/usr/local/bin/python
import urllib2
req = urllib2.Request('http://www.julesberman.info/factoids/batch.htm')
try:
    response = urllib2.urlopen(req)
except urllib2.HTTPError, e:
    print 'The server couldn\'t fulfill the request.'
    print 'Error code: ', e.code
except urllib2.URLError, e:
    print 'We failed to reach a server.'
    print 'Reason: ', e.reason
else:
    print urllib2.urlopen(req).read()
req = urllib2.Request('http://www.julesberman.info/factoids/xxxxx.htm')
try:
    response = urllib2.urlopen(req)
except urllib2.HTTPError, e:
    print 'The server couldn\'t fulfill the request.'
    print 'Error code: ', e.code
except urllib2.URLError, e:
    print 'We failed to reach a server.'
    print 'Reason: ', e.reason
else:
    print urllib2.urlopen(req).read()
exit
```

An excellent Web tutorial explaining the urllib2 module is available at

http://docs.python.org/dev/howto/urllib2.html

Ruby Script

```
#!/usr/local/bin/ruby
require `net/http'
Net::HTTP.start(`www.julesberman.info') do
  |http|
  response = http.get(`/factoids/batch.htm')
  if response.body[400,3].nil?
    puts "Code = #{response.code}"
    puts "Message = #{response.message}"
    response.each{|key,value| puts key + " " + value}
  else
    puts response.body[400,10000]
  end
  response = http.get(`/factoids/xxxxx.htm')
  if response.body[400,300].nil?
    puts "Code = #{response.code}"
    puts "Message = #{response.message}"
    response.each{|key,value| puts key + " " + value}
  else
    puts response.body[400,300]
  end
end
exit
```

For Ruby, the Net::HTTP module comes bundled with the Ruby interpreter, in the standard library. Another module, Net::FTP, requests files by FTP (File Transfer Protocol).

More information on Ruby's Net::HTTP module is available at

http://ruby-doc.org/stdlib/libdoc/net/http/rdoc/index.html

16.1.2 Analysis

Perl, Python, and Ruby use their own external modules for HTTP transactions. Each language's module has its own peculiar syntax. Still, the basic operation is the same: your script initiates an HTTP request for a Web file at a specific network address (the URL, or Uniform Resource Locator). A response is received, and the Web page is retrieved, if possible. Otherwise, the response will contain some information indicating why the page could not be retrieved.

In the example script, two Web pages were requested. The first is located at http://www.julesberman.info/factoids/batch.htm, and is a valid URL. The second is located at http://www.julesberman.info/factoids/xxxxx.htm, and is an invalid address.

You can see that, with a little effort, you can use this basic script to collect and examine a large number of Web pages. With a little more effort, you can write your own spider software that searches for Web addresses, and iteratively collects information from Web links within Web pages.

16.2 CGI Script for Searching the Neoplasm Classification

Here are the steps for using CGI scripts:

1. Get yourself a server account with access to a "public_html" directory and a "cgi-bin" subdirectory. This is usually accomplished by paying a commercial ISP (Internet Service Provider) for a Web account, or by asking your company or academic sponsor for an account. When you get your account, the provider will explain to you how you can deposit, via FTP, Web pages (that you create) onto the public_html directory. The provider will also explain how you can deposit your CGI scripts onto the cgi-bin subdirectory. He will also explain how you can assign settings to your CGI scripts that restrict access to certain sets of users. The provider will also tell you if there are limitations on the kinds of scripts permitted on the server (e.g., specific versions of a language might be required by the server, and the server may be set up for one language and not another).

2. Create a Web page that creates an HTML form. Almost every HTML book contains information about forms. Forms are HTML objects that accept user input and send the input to a designated server. Text boxes and radio buttons are commonly encountered form objects. They can be created in just a few lines of HTML code. You will put the Web page in your public_html directory. This Web page will be accessible to anyone in the world who happens to know the Web address of the HTML page. Your server manager will provide you with the Web address of your public_html directory, and the complete address of the Web page is simply the HTML file name appended to the directory address.

3. Create a script that sits in the cgi-bin subdirectory of a server, whose specific address is included in the form that you previously included in your Web page. When anyone viewing your Web page, enters information in the form, and submits the information (usually by clicking on a button in the form), the information will be sent to your server-side script and processed.

This describes the basic steps for a CGI script. With a little imagination, you can see the enormous power of this approach. The best thing about CGI is that you do not need to learn another language. You simply apply the programming skills you have already mastered.

The neoplasm taxonomy is an example of a medical nomenclature that is easy to parse, search, and produce an output in a preferred format. We can use the neoplasm

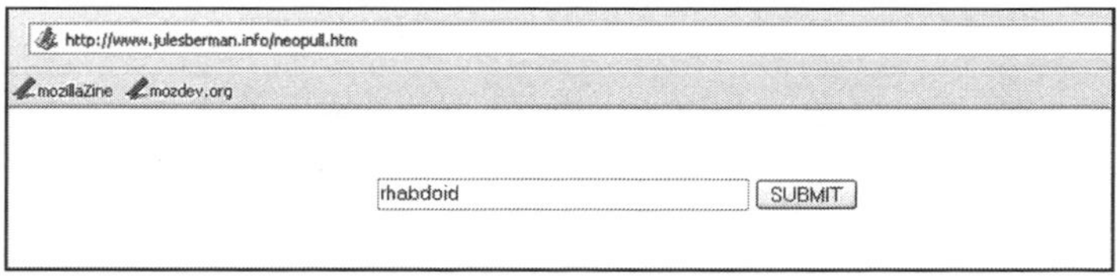

Figure 16.1 A Web page search box that will send a character string to your CGI script. In this case, we have entered the word "rhabdoid" into the text box.

taxonomy to search for neoplasm terms that match words and phrases submitted on a Web page. This will be our introductory CGI script.

16.2.1 Script Algorithm

1. Create a very simple Web page, consisting of a simple form containing a text box for user input (Figure 16.1). The form will contain the URL (Universal Resource Locator, or Web address) for the cgi-bin where your CGI script resides.
2. Upload the HTML document (your Web page) to the public_html directory on your Web server. Clients will send requests by entering information on the HTML document.
3. Create a script that you will upload to the cgi-bin of your server, which has the address specified in the Web page form (steps 1 and 2). The script will execute steps 4–8 when it receives a request from a client.
4. Capture the character string sent by the Web page, using command syntax specific to your preferred programming language, and place the text into a string object.
5. Print out the HTML header of the Web page that will be returned to the client (the user, sitting at a browser, somewhere on planet Earth, and looking at your Web page).
6. Process the text that the user sent to the CGI script. In this case, the information will be matched against every line in the neoself document, a 17+ megabyte (MB) collection of neoplasm terms that we previously created in Chapter 11. The neoself document must be deposited onto the server's cgi-bin.
7. Parse through every line of the neoself document. When a line that contains the string entered by the Web user is encountered, it is printed.
8. Print the HTML tags that mark the end of the Web page.

Perl Script

HTML text for client (requesting) web page:

```
<html>
<head>
<title>post</title>
</head>
```

```
<body>
<br><form name="sender" method="GET"
action="http://www.julesberman.info/cgi-bin/neopull.pl">
<br><center><input type="text" name="tx" size=38
maxlength=48 value="">
<input type="submit" name="bx" value="SUBMIT"></center>
</form>
<br><br>
</body>
</html>
```

```perl
#!/usr/local/bin/perl
print "Content-type: text/html\n\n";
$buffer = ($ENV{'QUERY_STRING'});
#read(STDIN, $buffer, $ENV{'CONTENT_LENGTH'});
print qq|<html><head><title></title></head><body>\n\n|;
if ($buffer =~ /Delete\+this\+and\+enter/)
   {
   print qq|\n <br><br>You didn't enter an neoplasm name in the
submit box, above |;
   print qq|\n <br><br></body></html> \n\n\n\n |;
   exit;
   }
if ($buffer =~ /^[a-zA-Z ]+$/)
   {
   print qq|\n <br><br>Only letters and spaces permitted.|;
   print qq|\n <br><br></body></html> \n\n\n\n |;
   exit;
   }
if ($buffer =~ /eval/i)
   {
   print qq|\n <br><br>No eval operators please |;
   print qq|\n <br><br></body></html> \n\n\n\n |;
   exit;
   }
$buffer =~ /tx\=([^&]+)&/;
$term = $1;
$term =~ s/%(..)/pack("c",hex($1))/ge;
print "Your entry was \<b\>$term\<\/b\>";
$term =~ s/omas/oma/o;
$term =~ s/tumo[u]*rs/tumor/o;
$term =~ s/neoplasms/neoplasm/o;
$term =~ s/kemias/kemia/o;
open (TERMS, "neoself");
$line = " ";
while ($line ne "")
   {
   $line = <TERMS>;
   if ($line =~ /$term/i)
      {
```

```
    $state = 1;
    print "\<br\>\n";
    $line =~ s/\|/\<br\>\n/g;
    print "\<br\>$line\n";
    }
  }
print qq|\n <br><br></body></html> \n\n\n\n |;
exit;
```

Python Script

HTML text for client (requesting) web page:

```
<html>
<head>
<title>post</title>
</head>
<body>
<br><form name="sender" method="GET"
action="http://www.julesberman.info/cgi-bin/neopull.py">
<br><center><input type="text" name="tx" size=38
maxlength=48 value="">
<input type="submit" name="bx" value="SUBMIT"></center>
</form>
<br><br>
</body>
</html>
```

```python
#!/usr/local/bin/python
import cgi, re, sys
import cgitb; cgitb.enable()
print "Content-type: text/html"
print
print "<html><head><title>Sample CGI Script</title></head><body>"
form = cgi.FieldStorage()
message = form.getvalue("tx", "(no message)")
term_check = re.search(r'[A-Za-z ]+$', message)
if not term_check:
  print "<br>Only alphabetic letters and spaces are permitted in
the query box"
  print "</body></html>"
  sys.exit()
print "<br>Your query term is " + message + "<br>"
in_text = open("neoself", "r")
for line in in_text:
  query_match = re.search(message, line)
  if query_match:
    line = re.sub(r'\|',"<br>", line)
    print "<br>" + line + "<br>"
exit
```

Ruby Script

HTML text for client (requesting) web page:

```
<html>
<head>
<title>post</title>
</head>
<body>
<br><form name="sender" method="GET"
action="http://www.julesberman.info/cgi-bin/neopull.rb">
<br><center><input type="text" name="tx" size=38
maxlength=48 value="">
<input type="submit" name="bx" value="SUBMIT"></center>
</form>
<br><br>
</body>
</html>
```

```
#!/usr/local/bin/ruby
print "Content-type: text/html\r\n\r\n"
print "<html><body></body></html>\r\n"
require 'cgi'
$SAFE = 1
cgi = CGI.new
query_term = cgi.params["tx"].to_s
if (query_term =~ /^[a-z\s]+$/i)
  query_term.untaint
else
  print "\<br\>Only alphabetic letters and spaces are permitted in
the query box\n"
  exit
end
print "\<br\>Your query term is #{query_term}\<br\>\<br\>\r\n"
text = File.open("neoself", "r")
text.each do
  |line|
  if (line =~ /#{query_term}/)
    line.gsub!(/\|/,"\<br\>\r\n")
    puts "\<br\>#{line}\<br\>"
  end
end
print "</body></html>\r\n"
exit;
```

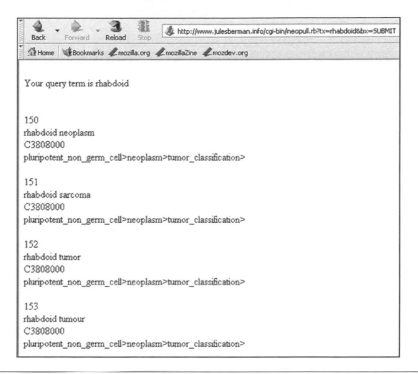

Figure 16.2 Output of search for "rhabdoid".

16.2.2 Analysis

In this case, the user entered the word "rhabdoid" into the Web page query box. The output immediately appears, as another Web page, in the same user's browser (Figure 16.2).

Notice that when the user pushes the "submit" button, all of the transmitted information appears in the browser's entry box, at the top of the Web page:

Exercises

1. Using Perl, Python, or Ruby, write a script that opens a Web page, searches for Web page addresses included in the Web page, and collects all of the contents of all the corresponding Web pages, putting the aggregate data into a single file.

2. Forms send information to server-side CGI scripts through either of two convenient message formats, "GET" or "POST." When a GET message is sent by an HTML form, the string containing the GET message appears at the browser's entry box. When a POST message is sent, the message does not appear in the browser. In the example provided in this chapter, the form sent a GET message. If the message were sent as a POST, the html form would appear as shown here:

```
<html>
<head>
<title>post</title>
</head>
<body>
<br><form name="sender" method="POST"
action="http://www.julesberman.info/cgi-bin/neopull.py">
<br><center><input type="text" name="tx" size=38
maxlength=48 value="">
<input type="submit" name="bx" value="SUBMIT"></center>
</form>
<br><br>
</body>
</html>
```

Using Perl, Python, or Ruby, revise your CGI script to accept a POST message (instead of a GET message).

3. In the example in this chapter, the output of the CGI script (the reply HTML Web page) contains the terms from the Neoplasm Classification that match the query term that was entered in the HTML page. Users who receive a reply to a form will expect to receive another form, just like the one they used originally, so that they can send another query. Using Perl, Python, or Ruby, revise the CGI script to insert a form (in the output html page, returned to the client) that will allow the user to submit another neoplasm term query.

4. In Chapter 10, Section 10.1, we developed a script that determines the lineage of organisms in Taxonomy.dat. Using Perl, Python, or Ruby, write an HTML page (for client input) and a CGI script that accepts the name of an organism as input and returns the ancestral lineage of the organism.

    Your project should look something like the following search engine page:

    http://www.julesberman.info/post.htm

5. In Chapter 3, Section 3.4, we showed how we can take an array of numbers and transform it into a simple bar graph. Using Perl, Python, or Ruby, write an HTML page that allows clients to input a series of comma-separated nonnegative numeric values, and a CGI script that returns a graph representing the numeric array.

17
IMAGE ANNOTATION

This chapter is written for people who need to annotate their photomicrographs in a manner that binds descriptive data to the image, so that

1. Collections of photomicrographs can be searched based on their descriptive content, or by their image content, or both.
2. Individual images can be sent to colleagues, and the person who receives the image can extract, from the image file, descriptive text that the sender included with the image.
3. After inserting text inside an image, the person who prepared the image can be certain that years later, after all the clinical and pathologic details associated with the image have been long forgotten, the image will still provide this information.
4. The data included in the image can be prepared in a standard form that is computer-parsable and understandable to software agents that search files on the Web.

Biomedical images have no value unless they are annotated with information that describes the image.

Important descriptors of an image might include

File information
Image-capture information
Image-format information
Specimen information
Patient information
Pathology information
Region-of-interest information

The easiest way of annotating an image is to compose a free-text description of your image and any other information you would like to add, such as your name, and adding the information as a Comment field in the header of the image file. The Comment will not alter the binary content of the image or the visual form of the image. When the file is copied, the header comment will be retained, and anyone receiving the image can read what you have added, using a simple script.

Professional curators may be held to a slightly higher standard. The Dublin Core is basic information designed by librarians to provide a minimal set of data to describe the contents of any electronic document. There are about 15 Dublin Core elements, including the name of the person who created the file, the date that the file was created, and the usage rights of the file. A professional image should contain all of the Dublin Core elements in its header. This chapter contains methods for inserting annotations into the popular formats for electronic images.

17.1 Inserting a Header Comment

Image files consist of binary information about the pixels (color spots) comprising the visual image, together with an image header that provides information explaining how the pixel data is organized (i.e., format-specific information). Within the image header are reserved blocks that can be enlarged with textual annotations. Using a reserved header block is an excellent way of conveying descriptive information within an image.

There are three common ways of inserting text data into the header of a binary image. You can add data to a comment field (sometimes called COM data), or your can add data to the two standard data fields used by camera manufacturers and commercial imaging applications (the EXIF and IPTC fields). We will show you simple scripts whereby you can enter and extract data contained in any of these three header blocks.

Additional information on EXIF is available at

http://www.exif.org/

Additional information on IPTC is available at

http://iptc.cms.apa.at/cms/site/index.html?channel=CH0099

17.1.1 Script Algorithm

1. Import the language-specific external module that supports modifications to image headers.
2. Create an image object, providing it with the filename of the image or images that you would like to modify. In this example script, you will need to substitute your own image files for any file names that appear in the script.
3. Add a comment to the image header, using the external module's command syntax.
4. Save the modified file. It is advisable to save the modified file as a second file, under a newly created filename, if you have not made backup copies of the original file.

Perl Script

Download the external module Image::MetaData::JPEG from the Perl packet manager (if ActiveState Perl is installed on your system, simply enter ppm as your command line and follow the instructions on the packet manager client).

```perl
#!/usr/local/bin/perl
use Image::MetaData::JPEG;
$file = new Image::MetaData::JPEG("saturn.jpg");
$file->add_comment("hello world");
$file->save("saturn2.jpg");
exit;
```

Python Script

```python
#!/usr/local/bin/python
def pngsave(im, file):
  from PIL import PngImagePlugin
  meta = PngImagePlugin.PngInfo()
  for k,v in im.info.iteritems():
      meta.add_text(k, v, 0)
  im.save(file, "PNG", pnginfo=meta)
from PIL import Image
image = Image.open("saturn.jpg")
image.save("saturn.png")
im = Image.open("saturn.png")
im.info["hello"] = "goodby"
im.info["now"] = "then"
pngsave(im, "saturn2.png")
exit
```

Ruby Script

Here is a Ruby script that inserts a comment into a PNG image and saves it as a JPEG image:

```ruby
#!/usr/local/bin/ruby
require 'RMagick'
include Magick
walnut = ImageList.new("neo1.png")
walnut.cur_image[:Comment] = "Hello World"
walnut_copy = ImageList.new
walnut_copy = walnut.cur_image.copy
walnut_copy.write("out.jpg")
exit
```

Here is a Ruby script that inserts a Comment and a Label into the JPEG header:

```ruby
#!/usr/local/bin/ruby
require 'RMagick'
include Magick
walnut = ImageList.new("c\:\\ftp\\rb\\CT4192~1.JPG")
walnut.cur_image[:Label] = "Salutations"
walnut.cur_image[:Comment] = "Hello World"
#walnut.properties{|name, value| print "#{name} #{value}\n"}
walnut_copy = ImageList.new
walnut_copy = walnut.cur_image.copy
exit
```

This Ruby script inserts a header into a TIFF file:

```
#!/usr/local/bin/ruby
require 'RMagick'
include Magick
tissue = ImageList.new("submu_bw.tif")
tissue.cur_image[:Comment] = "Hello world"
tissue_copy = ImageList.new
tissue_copy = tissue.cur_image.copy
tissue_copy.write("submu_2.tif")
exit
```

17.1.2 Analysis

After a comment has been inserted into an image header, it can be modified at a later date. Modifications should be dated and recorded within the image header.

17.2 Extracting the Header Comment in a JPEG Image File

Once you have prepared a file with your own comments inserted into the image header, you will need a way to extract that information.

17.2.1 Script Algorithm

1. Import any necessary image modules.
2. Create an image object, providing the filename of the image of interest. In this example script, you will need to substitute your own image files for any file names that appear in the script.
3. Using the module operator that gets header comments, fetch the comments and output the returned text to the monitor.

Perl Script

```
#!/usr/local/bin/perl
use Image::MetaData::JPEG;
$file = new Image::MetaData::JPEG("saturn2.jpg");
print join("", $file->get_comments());
exit;
```

Python Script

```
#!/usr/local/bin/python
from PIL import Image
im = Image.open("saturn2.png")
print im.info
exit
```

Ruby Script

To extract just the Comment section:

```
#!/usr/local/bin/ruby
require 'RMagick'
include Magick
walnut = ImageList.new("out.jpg")
print walnut.properties.fetch("Comment")
exit
```

The same method works for most image formats:

```
#!/usr/local/bin/ruby
require 'RMagick'
include Magick
tissue = ImageList.new("submu_2.tif")
print tissue.properties['Comment']
exit
```

To extract all of the annotated properties, including the Comment section:

```
#!/usr/local/bin/ruby
require 'RMagick'
include Magick
walnut = ImageList.new("out.jpg")
walnut.properties{|name, value|puts "#{name} #{value}"}
exit
```

The same script words for PNG images:

```
#!/usr/local/bin/ruby
require 'RMagick'
include Magick
img = Magick::Image::read("c\:\\ftp\\py\\mes.png").first
img.properties{|name,value| puts "#{name} #{value}"}
exit
```

17.2.2 Analysis

Comments can be lost by inadvertently overwriting an old comment when a new comment is inserted. It is good practice to convert a copy of an image (rather than the original image) when changing image formats. When inserting comment text into an image, make sure there is no preexisting comment that may be overwritten by the new comment. If there is a preexisting comment, you can concatenate it with the new comment, and insert the combined text as your new header text.

Reading embedded image comments is a simple operation, and scripts should not avoid checking on header content.

17.3 Inserting IPTC Annotations

Photographers who license their images often insert copyright and contact information in IPTC (International Press Telecommunications Council) headers. As a biomedical professional, you will not need to insert IPTC data. You can insert copyright and contact information in the comment section, as Dublin Core tags (see Chapter 18, Section 18.2).

17.4 Extracting Comment, EXIF, and IPTC Annotations

Whenever you receive an image that you intend to use for patient care, research, or any medical activity, you should examine the header contents. Luckily, it is much easier to extract the textual contents of an image than to insert data. Image Magick's "Identify" command extracts header data inserted in Comment, EXIF (used by many digital cameras), or IPTC formats.

17.4.1 Script Algorithm

1. Use a system call to invoke Image Magick's Identify command.
2. Modify the Identify command to produce a verbose output. In this example script, you will need to substitute your own image files for any file names that appear in the script.
3. Redirect the output to a text file.

Perl Script

```
#!/usr/local/bin/perl
system("Identify -verbose c\:\\ftp\\3320_out.jpg \>myimage.txt");
exit;
```

Python Script

```
#!/usr/local/bin/python
import os
os.system("Identify -verbose c:/ftp/3320.jpg >myimage.txt");
exit
```

Ruby Script

```
#!/usr/local/bin/ruby
system("Identify -verbose c:/ftp/3320.jpg >myimage.txt")
exit
```

17.4.2 Analysis

Although scripting languages have their own interfaces to ImageMagick, and ImageMagick has specific methods for extracting the information in the Comment,

```
Format: JPEG (Joint Photographic Experts Group JFIF format)
Class: PseudoClass
Geometry: 1342x1177
Type: Grayscale
Endianess: Undefined
Colorspace: Gray
Channel depth:
  Gray: 8-bit
Channel statistics:
  Gray:
    Min: 0 (0)
    Max: 255 (1)
    Mean: 178.121 (0.698515)
    Standard deviation: 70.9532 (0.278248)
Colors: 256
Histogram:
Rendering intent: Undefined
Resolution: 96x96
Units: PixelsPerInch
Filesize: 754.451kb
Interlace: None
Background color: white
Border color: #DFDFDF
Matte color: grey74
Page geometry: 1342x1177+0+0
Dispose: Undefined
Iterations: 0
Compression: JPEG
Quality: 99
Orientation: Undefined
Comment:
<?xml version="1.0" encoding="UTF-8"?>
<rdf:RDF
    xmlns:rdf="http://www.w3.org/1999/02/22-rdf-syntax-ns#"
    xmlns:dc="http://purl.org/dc/elements/1.1/">
  <rdf:Description rdf:about="http://www.julesberman.info/">
    <dc:creator>Jules J. Berman</dc:creator>
    <dc:title>Methods in Medical Informatics</dc:title>
    <dc:description>
    Medical Informatics methods and algorithms in Perl, Python, and Ruby
    </dc:description>
    <dc:date>2010</dc:date>
  </rdf:Description>
</rdf:RDF>
JPEG-Colorspace: 1
JPEG-Sampling-factors: 1x1
Signature: aaaaa08e4f76c531647b6e0a5ea6de23e995dca63e69b5ac4946918fa72a248b
Tainted: False
User time: 0.078u
Elapsed time: 0:01
Pixels per second: 19.0118mb
Version: ImageMagick 6.2.9 08/11/06 Q8 http://www.imagemagick.org
```

Figure 17.1 The verbose output of ImageMagick's "Identify" command for the 3320_out.jpg image file, showing all the header information, with the exception of the "Histogram" listing, which is many lines in length. Near the bottom is the "Comment" that we inserted into the header, in Chapter 18, Section 18.3.

EXIF, and IPTC header sections, there really is no advantage in using any of these language-specific techniques. If you have ImageMagick installed on your computer, a (language-independent) system call to the "Identify" command will collect any and all information that is in the image header. A partial listing of the output is shown in Figure 17.1.

When EXIF or IPTC data is included in an image header, the header output will contain the word "Profile", followed by "8BIM" to mark the beginning of IPTC information; and APP1, to mark the beginning of EXIF data. You may occasionally see "Profile", followed by APP12. APP12 was used in early digital cameras.

17.5 Dealing with DICOM

In the field of biomedicine, DICOM (Digital Imaging and Communications in Medicine) is the format currently used for radiologic images and in hospital-based picture archiving and communication systems (PACS). DICOM was developed over several decades. DICOM is very complex, using a model for data storage that is unlike any other image file format. The DICOM standard includes transfer and communication protocols, used to negotiate the exchange of information between radiologic devices or different parts of a single device (e.g., between CT machine and CT workstation). Very few people outside the radiology device field fully understand the DICOM standard.

For most purposes, you will likely be using images saved in the JPEG format, which is the favored format for millions (possibly billions) of Web images. Most digital cameras save images as JPEG files, and, as we have seen, programming libraries have access to free modules that modify or convert JPEG images. If you work in a medical center or have colleagues who use medical center PACS software, you may need to deal with DICOM. Unless you specialize in radiologic images, you will probably want to convert DICOM images to JPEG or some other popular image format, that can be easily handled with Perl, Python, or Ruby. If you need to deliver images in DICOM format to your medical colleagues, it is easy enough to convert your JPEG images back to DICOM, as necessary.

17.6 Finding DICOM Images

You can find many DICOM images by anonymous ftp at

ftp://ftp.erl.wustl.edu/pub/dicom/images/version3/RSNA95/

These images can be used as practice files for your own scripts.

DICOM has a header that can be extracted from the DICOM image file; it contains textual descriptive information about the image (Figure 17.2).

As you can see, most of the header information in a DICOM file is radiology oriented.

```
0010,0020,Patient ID=GE1115
0010,0030,Patient Date of Birth=
0010,0040,Patient Sex=M
0010,1010,Patient Age=
0010,1030,Patient Weight=8240
0018,0000,Acquisition Group Length=172
0018,0010,Contrast/Bolus Agent=15    GAD-P
0018,0020,Scanning Sequence=GR
0018,0021,Sequence Variant=
0018,0022,Scan Options=
0018,0023,MR Acquisition Type=
0018,0050,Slice Thickness=0.000000
0018,0080,Repetition Time [TR, ms]=48.000000
0018,0081,Echo Time [TE, ms]=0.000000
0018,0082,Inversion Time=0.000000
0018,0087,Magnetic Field Strength=15000
0018,0091,Echo Train Length=
0018,1060,Trigger Time=-1.000000
0018,5100,Patient Position=HFS
0020,0000,Relationship Group Length=282
0020,000D,Study Instance UID=1.2.840.113674.1115.261.200.
0020,000E,Series Instance UID=1.2.840.113674.1115.261.178.300.
0020,0010,Study ID=29172
0020,0011,Series Number=101
0020,0012,Acquisition Number=
0020,0013,Image Number=0
0020,0020,Patient Orientation=A\F
0020,0032,Image Position Patient=2.750600\107.312401\96.091698
0020,0037,Image Orientation (Patient)=0.000000\-1.000000\0.000000\0.000000\0.000000\-1.000
0020,0052,Frame of Reference UID=
0020,0060,Laterality=
0020,1040,Position Reference=
0020,1041,Slice Location=0.000000
0020,4000,Image Comments=
```

Figure 17.2 Part of a typical DICOM header.

17.7 DICOM-to-JPEG Conversion

ezDICOM (copyright 2002, Wolfgang Krug and Chris Rorden) is a medical viewer for DICOM images. It is distributed along with dcm2jpg, a command-line application that can convert DICOM images into standard bitmap file formats (JPEG, PNG, and BMP). In addition, it will convert a DICOM image to its textual header information.

You can download the freeware DICOM to JPEG converter from

http://www.cabiatl.com/mricro/ezdicom/

The downloaded zipped file is ezdicom.zip. When you de-archive the zip file, save the dcm2jpg.exe file. The dcm2jpg.exe will convert a DICOM file (on the command line) to a JPEG file. When you rename the .exe file, you change the default conversion behavior:

dcm2jpg.exe 218,112 bytes—converts to DICOM to JPEG format
dcm2bmp.exe 218,112 bytes—converts to DICOM to BMP format
dcm2txt.exe 218,112 bytes—extracts the text header from a DICOM file

To extract the text header from a DICOM file, invoke the following command line in the subdirectory containing dcm2txt.exe:

c:\ftp>dcm2txt.exe c:\ftp\1.dcm
1 Creating: c:\ftp\1.txt

This creates a text file that looks something like the sample header figure. Save this text file. We can insert it into the Comment section of a JPEG image. The dcm2jpg.exe software can be executed from within a Perl, Python, or Ruby script.

17.7.1 Script Algorithm

1. From the script, make a system call to the external program dcm2jpg.exe, supplying the DICOM file name, and its full path. In this example script, you will need to substitute your own image files for any file names that appear in the script. This will create a JPEG file in the same directory as the DICOM file.
2. From the script, make a system call to the external program dcm2txt.exe, supplying the DICOM file name and its full path. This will create a text file of the DICOM header in the same directory as the DICOM file.

Perl Script

```
#!/usr/local/bin/perl
system("dcm2jpg.exe c\:\\ftp\\p1\\1.dcm");
system("dcm2txt.exe c\:\\ftp\\p1\\1.dcm");
exit;
```

Python Script

```
#!/usr/local/bin/python
import os
os.system("dcm2jpg.exe c:/ftp/py/1.dcm")
os.system("dcm2txt.exe c:/ftp/py/1.dcm")
exit;
```

Ruby Script

```
#!/usr/local/bin/ruby
system("dcm2jpg.exe c:/ftp/py/1.dcm")
system("dcm2txt.exe c:/ftp/py/1.dcm")
exit
```

17.7.2 Analysis

The script produces two files: 1.jpg and 1.txt. The JPEG file lacks the clinical header that was contained in the DICOM file, but this information is now available to us in the 1.txt file. The textual DICOM header can be inserted back into a JPEG file, using the methods described earlier in this chapter.

Exercises

1. Use Perl, Python, or Ruby to add a free-text comment into a JPEG file. Convert your file to another file format using your favorite image software. Use Perl, Python, or Ruby to extract your comment from the image file in its new format. Was the comment preserved when the file format was changed? Write a short script that extracts your comment from your JPEG image, converts the JPEG image to another format, checks to see if the comment was preserved, and reinserts the comment into the reformatted image if the comment was not saved during the reformatting step.

2. Using Perl, Python, or Ruby, write a script that parses a collection of at least 10 images, searching for EXIF, IPTC, and COM comments, and prepares a file listing each image (by image filename), along with all of the extracted metadata for each image.

3. Create a subdirectory, and fill it with 10 images of your choice, in TIFF, PDF, GIF, or PNG formats. Using Perl, Python, or Ruby, write a script that will insert your name, and the date into the header for each image, yielding files of the same name and format as the original image.

4. Using Perl, Python, or Ruby, convert each of the images from Exercise 3 to a JPEG image, carrying the same header.

5. Using Perl, Python, or Ruby, write a script that will produce an HTML document that displays each of the 10 images, with the header of each image appearing directly below the image on the displayed page.

6. Write a script, using Perl, Python, or Ruby, that converts a DICOM image to a JPEG image, extracts the text header from the DICOM image, and inserts the DICOM header into the comment section of the JPEG image.
7. Write a new script, in Perl, Python, or Ruby, that opens the directory containing at least 10 DICOM files, and converts each of the DICOM files to a JPEG file.

18
DESCRIBING DATA WITH DATA, USING XML

The importance of XML as a data-organizing tool cannot be overstated. As a data-organizing technology, it is as important as the invention of written language (circa 3000 BC) or the mass-printed book (circa 1450 AD). At its simplest, XML is a method for marking up files so that every piece of data is surrounded by bracketed text that describes the piece of data (e.g., <number>5</number>). Markup allows us to convey any message as XML (a pathology report, a radiology image, a genome database, a software program, an e-mail).

XML markup tags are sets of alphanumeric descriptors enclosed by angle brackets. Each tag is repeated at the beginning and end of the data element, the ending tag demarcated by the slash character "/".

The following are examples of XML markup:

```
<name_of_patient>John Public</name_of_patient>
<age_of_patient>25 years</age_of_patient>
<gender_of_ patient>Male</gender_of_ patient>
<birthdate_of_patient>January 1, 1954</birthdate_of_patient>
```

The same data could have been nested as children of a father tag:

```
<patient>
<name>John Public</name>
<age>25 years</age>
<gender>Male</gender>
<birthdate>January 1, 1954</birthdate>
</patient>
```

A file that contains XML markup is considered an XML file only if it is well formed. That is, it must have a proper XML header; it must consist of text in a readable form (typically, the simple letters and punctuation found on a keyboard), and it must follow the general rules for using tagging data. The header can vary somewhat, but it usually looks something like: <?xml version="1.0"?>. Tags must have a certain form (e.g., spaces are not permitted within a tag), and tags must be properly nested (i.e., no overlapping). For example, <chapter><chapter_title>Informaticians love XML</chapter_title></chapter> is nicely nested XML. <chapter><chapter_title>Pathologists love XML</chapter></chapter_title> is improperly nested.

Because XML follows a strict syntax, it is relatively easy to write parsing scripts that extract the data values and descriptors in XML files. It is also easy to write parsing scripts that detect violations of XML syntax rules.

18.1 Parsing XML

In Chapter 11, we wrote an XML parser for the neoplasm taxonomy. While parsing the file, our script automatically checked to determine that the file is well-formed XML (i.e., if it conforms to the rules of XML syntax). Had there been any nonconforming lines or characters anywhere in the 10+ megabyte (MB) neoplasm taxonomy file, our script would have indicated the specific lines in the file where a syntactic error occurred. Let us write a script whose only purpose is to check XML documents for proper syntax.

18.1.1 Script Algorithm

1. Import an XML parsing module.
2. Create a new parser object.
3. Using a (parsing) method available in the parsing module, provide the method with the name of the file you wish to parse.
4. The parsing module will send a message to your screen if any parts of the file are not well formed.

Perl Script

```
#!/usr/local/bin/perl
use XML::Parser;
my $parser = XML::Parser->new( Handlers => {
Init => \&handle_doc_start,
Final => \&handle_doc_end,
});
$parser -> parsefile("c\:\\ftp\\neocl.xml");
sub handle_doc_start
{
print "\nBeginning to parse file now\n";
}
sub handle_doc_end
{
print "\nFinished. $file is a well-formed XML File.\n";
}
exit;
```

Sample output:

```
Beginning to parse file now
mismatched tag at line 138649, column 2, byte 10228415
at C:/Perl/lib/XML/Parser.pm line 187
```

Python Script

```
#!/usr/local/bin/python
import xml.sax
import pprint
parser = xml.sax.make_parser( )
parser.parse('C:\\ftp\\neocl.xml')
exit
```

Sample output:

```
C:\ftp\py>python rexml.py
Traceback (most recent call last):
File "rexml.py", line 7, in <module>
parser.parse('C:\\ftp\\neocl.xml')
File "C:\Python25\lib\xml\sax\expatreader.py", line 107, in parse
xmlreader.IncrementalParser.parse(self, source)
File "C:\Python25\lib\xml\sax\xmlreader.py", line 123, in parse
self.feed(buffer)
File "C:\Python25\lib\xml\sax\expatreader.py", line 211, in feed
self._err_handler.fatalError(exc)
File "C:\Python25\lib\xml\sax\handler.py", line 38, in fatalError
raise exception
xml.sax._exceptions.SAXParseException: C:\ftp\neocl.xml:138654:2:
mismatched tag
```

Ruby Script

```
#!/usr/local/bin/ruby
require 'rexml/document'
require 'rexml/streamlistener'
include REXML
class Listener
  include StreamListenerend
listener = Listener.new
parser = Parsers::StreamParser.new(File.new("c:/ftp/neocl.xml"),
listener)
parser.parse
exit
```

Sample output:

```
C:\ftp\rb>ruby rexml.rb
c:/ruby/lib/ruby/1.8/rexml/parsers/baseparser.rb:315:in `pull':
Missing end tag
for 'stage' (got "unclassified") (REXML::ParseException)
Line: 138654
Position: 10228801
Last 80 unconsumed characters:
from c:/ruby/lib/ruby/1.8/rexml/parsers/streamparser.rb:16:in `parse'
from rexml.rb:12
```

18.1.2 Analysis

This script takes just a few lines of code, and parses XML files very quickly. The script determines whether the XML file is well formed. For this example, I deliberately opened the 10+ MB neocl.xml file, and created a syntax error by removing the end of the stage tag a few lines from the end of the file. The script found the error and reported the file location where the error occurred. Syntax errors, when they occur, are always detected.

There are basically two types of XML parsing methods: stream methods (such as our script), and DOM (Document Object Model) methods.

Stream methods parse through the file, much like any text parser, until an XML "event" occurs (such as an encounter with the beginning of an XML tag, or the end of an XML tag). When an event occurs, information is collected that must be reconciled with subsequent events (e.g., every tag must have an end tag, and child tags must end before the parent tag ends). The streaming parsers permit users to add additional commands to be executed during an event.

DOM parsers build a model of the XML structure (i.e., all the XML objects and their relationships to each other). DOMs allow us to use the relationships among XML objects in applications. The drawback of DOM parsers is that iterations up and down the relational model, as the XML document is parsed, slow the script. A large XML document (many megabytes) with a complex XML structure can take a very long time to parse.

Because I tend to use big XML documents, with many child elements, that have long lineages, I use stream parsers exclusively. I suspect that healthcare workers, who use large XML data sets, will tend to rely on stream parsers.

18.1.3 Resource Description Framework (RDF)

The text in this section is an excerpt from an open access document (Berman, J.J., Moore, G.W. Implementing an RDF Schema for Pathology Images, from the Association for Pathology Informatics, APIII, Pittsburgh, PA, September 10, 2007. http://www.julesberman.info/spec2img.htm.)

RDF is a variant of XML, and conforms to the same tagging syntax as XML. The key difference between RDF and XML is that paired data and metadata (the essence of XML) is always bound to an identified object, forming a data "triple."

Triples consist of specified subject, metadata, and data, in that sequence.

Examples of triples found in a medical data set:

"Mr. Rheeus" "blood glucose level" "77"

The data is the number, "77". The metadata is the descriptor, "blood glucose level". The specific object is "Mr. Rheeus".

```
<Description>
<Description_object>Mr. Rheeus</Description_object>
<Blood_glucose_level>77</Blood_glucose_level>
</Description>
```

This kind of XML statement demonstrates the sharp distinction between data–metadata pairs, and a meaningful assertion. A data–metadata pair does not make a meaningful statement. When the data–metadata pair is bound to an object (the thing that the data–metadata pair is "about"), then you have a meaningful assertion. In this example, we are saying that there is a blood glucose level of 77 that belongs to Mr. Rheeus. Triples can be parsed, collected, combined with other triples (bound to the same object) collected from diverse data sources, repackaged (in new RDF documents), searched, and analyzed.

As you might expect, RDF has its own syntax for expressing triples:

```
<rdf:Description rdf:about="http://www.patient_info.com/lab.htm#Mr_Rheeus">
<lab:Blood_glucose_level>77</lab:Blood_glucose_level>
</rdf:Description>
```

In RDF, objects are specified by a Web address, or some identifier that uniquely distinguishes the object from all other objects.

An RDF document might look something like this:

```
<?xml version="1.0" encoding="UTF-8"?>
<rdf:RDF
xmlns:rdf="http://www.w3.org/1999/02/22-rdf-syntax-ns#"
xmlns:lab="http://lab.org/elements/">
<rdf:Description rdf:about="http://www.patient_info.com/lab.htm#Mr_Rheeus">
<lab:Blood_glucose_level>77</lab:Blood_glucose_level>
</rdf:Description>
</rdf:RDF>
```

An important feature of RDF is its ability to fully describe all of the metadata included in an RDF document, and to provide a class hierarchy for metadata. The descriptions and hierarchical organization of metadata are provided in external documents, called Schemas, that are referenced from within RDF documents that use the metadata tags described in the Schema.

In the short example here, the "Blood_glucose_level" tag is described in an external document, at "http://lab.org/elements/"

RDF is the syntax and logic underlying the semantic Web, and every serious informatician must learn to use RDF.

For the purposes of this book, we will only be examining a very specific example of RDF annotation, the Dublin Core.

```
- <rdf:RDF>
  - <rdf:Description rdf:about="http://www.julesberman.info/">
      <dc:creator> Jules J. Berman</dc:creator>
      <dc:title> Methods in Medical Informatics</dc:title>
  - <dc:description>
      Medical Informatics methods and algorithms in Perl, Python, and Ruby
    </dc:description>
      <dc:date>2010</dc:date>
    </rdf:Description>
  </rdf:RDF>
```

Figure 18.1   A simple RDF file, containing the Dublin Core elements: creator, title, description, and date.

18.2 Dublin Core Metadata

The Dublin Core consists of about 15 data elements, selected by a group of librarians, that specify the kind of file information a librarian might use to describe a file, index the described file, and retrieve files based on included information.

There are many publicly available documents that describe the Dublin Core elements:

http://www.ietf.org/rfc/rfc2731.txt

The Dublin Core elements can be inserted into HTML documents, simple XML documents, or RDF documents. A public document explains exactly how the Dublin Core elements can be used in these file formats:

http://dublincore.org/documents/usageguide/#rdfxml

An example of a simple, and shortened, Dublin Core file description in RDF format is shown in Figure 18.1:

Because RDF is a dialect of XML, we can parse RDF files with the same scripts that parse XML files. Because XML (and RDF) are ASCII files, they can be inserted into the header sections of image files. When Dublin Core RDF is inserted into an image file, it can be easily extracted and used to identify the file, and the individual Dublin Core elements can be combined with Dublin Core elements from other files to organize a wide range of data sources.

18.3 Insert an RDF Document into an Image File

It is easy to insert an RDF document into the header of a JPEG image file, and it is just as easy to extract the RDF triples.

18.3.1 Script Algorithm

1. Prepare your RDF document. In this case, we will use the RDF file containing a few Dublin Core elements, available at

   http://www.julesberman.info/book/rdf_desc.xml

2. Open an image file. In this case, we use the JPEG file/3320.jpg, available at

   http://www.julesberman.info/book/3320.jpg

3. Insert the RDF document into the Comment section of the JPEG header.
4. Save the file.
5. Extract the header comments.

Perl Script

The Perl script requires the freely available open source module, Image::MetaData::-JPEG. You can download this module from CPAN (Comprehensive Perl Archive Network, www.cpan.org).

```perl
#!/usr/local/bin/perl
use Image::MetaData::JPEG;
my $filename = "3320.jpg"; #comment:your filename here
my $file = new Image::MetaData::JPEG($filename);
die 'Error: ' . Image::MetaData::JPEG::Error() unless $file;
print "Description of JPEG file\n";
print $file->get_description();
print "\n\nRDF Annotations to JPEG file\n\n";
open (TEXT, "rdf_desc.xml")||die"cannot";
$line = " ";
while ($line ne "")
    {
    $line = <TEXT>;
    $file->add_comment($line);
    }
unlink $filename;
$file->save($filename);
my $file = new Image::MetaData::JPEG($filename);
my @comments = $file->get_comments();
print join("",@comments);
exit;
```

Python Script

```python
#!/usr/local/bin/python
def pngsave(im, file):
  from PIL import PngImagePlugin
  meta = PngImagePlugin.PngInfo()
```

```
for k,v in im.info.iteritems():
    meta.add_text(k, v, 0)
  im.save(file, "PNG", pnginfo=meta)
from PIL import Image
image = Image.open("c:/ftp/3320.jpg")
image.save("c:/ftp/3320.png")
rdf_file = open("c:/ftp/rdf_desc.xml", "rb")
description = rdf_file.read()
rdf_file.close()
im = Image.open("c:/ftp/3320.png")
im.info["description"] = description
pngsave(im, "c:/ftp/3320.png")
exit
```

Ruby Script

```
#!/usr/local/bin/ruby
require 'RMagick'
include Magick
text = IO.read("c:/ftp/rdf_desc.xml")
orig_image = ImageList.new("c:/ftp/3320.jpg")
orig_image.cur_image[:Comment] = text
copy_image = ImageList.new
copy_image = orig_image.cur_image.copy
copy_image.write("c:/ftp/rb/3320_out.JPG")
exit
```

18.3.2 Analysis

When you include Dublin Core elements in your image headers, you accomplish several very important goals at once:

1. You provide your colleagues with important descriptive information about the image.
2. You provide indexing services and search engines with information that they can extract, from your Web-residing images, that permits others to find your images.
3. If you provide copyright information and language that fully explains the rights of the creator and the user, you can ensure that anyone who acquires your image will have the information they need to use your intellectual property in a responsible and legal manner.
4. You turn your image into a mini-database, that can be integrated with other database files.

18.4 Insert an Image File into an RDF Document

Though we distinguish text files from binary files, all files are actually binary files. Sequential bytes of 8 bits are converted to ASCII equivalents, and if the ASCII equivalents are alphanumeric, we call the file a text file. If the ASCII values of 8-bit sequential file chunks are nonalphanumeric, we call the files binary files.

Standard format image files are always binary files. Because RDF syntax is a pure ASCII file format, image binaries cannot be directly pasted into an RDF document. However, binary files can be interconverted to and from ASCII format, using a simple software utility.

18.4.1 Script Algorithm

1. Call the external Base64 module.
2. Use any image file. In the example, we use 3320.jpg, available for download at

   http://www.julesberman.info/book/3320.jpg.

3. Put the entire contents of the image file into a string variable.
4. Encode the contents of the image file into base64, using the encoding method from the external module.
5. Open the RDF file. In this example, we will use the rdf_desc.xml file, available at

   http://www.julesberman.info/book/rdf_desc.xml

6. Split the file on the <dc:description> tag, and put the base64-encoded string into this tagged data section.
7. Mark the base64 text with "BEGIN" and "END."
8. Put the modified contents of the rdf_desc.xml file, now containing the base64 representation of the image file, into a new file, named rdf_image.xml.

Perl Script

```perl
#!/usr/local/bin/perl
use MIME::Base64::Perl;
open (JPGBIN,"c\:\\ftp\\3320.jpg")||die"cannot";
binmode JPGBIN;
$/ = undef;
$image_string = <JPGBIN>;
close JPGBIN;
$encoded = encode_base64($image_string);
open (RDF, "c:\\ftp\\rdf_desc.xml")||die"cannot";
$rdf_string = <RDF>;
print $rdf_string;
$rdf_string =~ /\<dc\:description\>/;
$contents = $` . $& . "BEGIN\n" . $encoded . "END\n" . $';
open(OUT,">c\:\\ftp\\rdf_image.xml");
print OUT $contents;
exit;
```

Python Script

```
#!/usr/local/bin/python
import base64, re
image_file = open("c:/ftp/3320.jpg", "rb")
image_string = image_file.read()
image_file.close()
contents = ""
encoded = base64.encodestring(image_string)
rdf_file = open("c:/ftp/rdf_desc.xml", "r")
rdf_string = rdf_file.read()
rdf_file.close()
rdflist = re.split(r'dc:description>', rdf_string)
contents = rdflist[0] + "dc:description>BEGIN\n" + \
encoded + "END\n" + rdflist[1] + "dc:description" + rdflist[2]
rdf_out = open("c:/ftp/rdf_image.xml", "w")
print>>rdf_out, contents
exit
```

Ruby Script

```
#!/usr/local/bin/ruby
require 'base64'
image_string = File.new("c:/ftp/3320.jpg").binmode.read
encoded = Base64.encode64(image_string)
rdf_string = File.open("c:/ftp/rdf_desc.xml", "r").read
rdf_string =~ /\<dc\:description\>/
contents = $` + $& + "BEGIN\n" + encoded + "END\n" + $'
rdf_out = File.open("c:/ftp/rdf_image.xml", "w")
rdf_out.print contents
exit
```

18.4.2 Analysis

The abbreviated output is shown in Figure 18.2.

The full file exceeds a megabyte in length. The central section of the base64 image block is removed, to permit us to see the structure of the output file.

The sample script is not particularly robust. It requires the presence of a Dublin Core description tag appearing in an exact format (i.e., <dc:description>). Otherwise, the script would just fail. The script inelegantly shoves the base64 representation of the binary image data into the Dublin Core description field. If this were a real RDF implementation, you would prepare a specific RDF tag for the base64 data, and you would prepare an external Schema document that defined the tag and its properties.

The script shows us that RDF files can hold binary data files (represented as base64 ASCII strings). There may be instances when you might prefer to insert an image file into an RDF document, rather than inserting an RDF document into an image file.

```
<?xml version="1.0" encoding="UTF-8"?>
<rdf:RDF
     xmlns:rdf="http://www.w3.org/1999/02/22-rdf-syntax-ns#"
     xmlns:dc="http://purl.org/dc/elements/1.1/">
<rdf:Description rdf:about="http://www.julesberman.info/">
     <dc:creator>Jules J. Berman</dc:creator>
     <dc:title>Methods in Medical Informatics</dc:title>
     <dc:description>BEGIN
/9j/4AAQSkZJRgABAQEAYABgAAD/2wBDAAEBAQEBAQEBAQEBAQEBAQEB
AQEBAQEBAQEBAQEBAQICAQECAQEBAgICAgICAgICAQICAgICAgICAgL/
wAALCASZBT4BAREA/8QAHwAAAQUBAQEBAQEAAAAAAAAAAAECAwQFBgcICQoL
/8QAtRAAAgEDAwIEAwUFBAQAAAF9AQIDAAQRBRIhMUEGE1FhByJxFDKBkaEI
IOKxwRVSOfAkM2JyggkKFhcYGRolJicoKSo0NTY3ODk6Q0RFRkdISUpTVFVW
V1hZWmNkZWZnaGlqc3R1dnd4eXqDhIWGh4iJipKTlJWWl5iZmqKjpKWmp6ip
qrKztLW2t7i5usLDxMXGx8jJytLT1NXW19jZ2uHi4+Tl5ufo6erx8vP09fb3
+Pn6/9oACAEBAAA/AP7VX1O8GoXbNf6pHaW8mo28Bku9UWOKae+8aRrNJJLq
ESyIrJFxLIEVYQGdEXiymrRuswk1Ev5NpNcFV1KAm8WKNnMC7/HWOFgHC5HJ
YcHJFZc9w99c3N28+oQRxWVxNBq2mQxqpezg7I/G65G+6uTnOYAfcQLZtr
8RQskd9dXs8sMs1yJdZtI2itibuSZ4TbeMnZpA15dMNsTyZZQiOVRRmLe6gv
/H3f6pOAwMUR1bUI/Mth1hLRXQJDJuBPLc9aq22rXOd7LOWu4y8TwIsupX7R
BZAUCvvG9OQeG5Y/NxkNWzNd3NwW2sCtraSXcixarqSvmCMyZBS8HzE4AHrjI
.
.
.
.  [LOTS MORE HERE]
.
.
j1HOb+RqY/dP/Xs//oFOp/rG/wCuD/8AoBq52k/69JP/AEUaqD/UJ/yD/wBB
NNk/1Kf9dB/6EtaF39yH/riP5CsJ+sX/AF1T+Zq8PuN9f8KkT/j5h+o/9CFM
P/Hxcf7p/k1RTf68/+5nVqdH92b6P/6A1ZOn+tT/AHH/ABb/7LTbz/j3P8A
rjc/+gGsO/49o/90fIyNcXqH3j+P8jWGn31/31/mK2bb/j1/GL/dxdp/rmf
6UqdE+o/ktLs/wDLH/dn/rV5f7Yf+ub/APLQdn/q4fOC/D+Qpp/17/6EKtX
matL9+b/AK5SQ/wDoNP8A4W/4F//MOQffH+8n86jf7x/D+Qpp/17/9cT+6EktX
HWL/AK4j/wBArKX/wBArFtn9Zv5Vdvx7x/D+Qqxb/3Z/Gvebov+v8A6FEtX
/wDLO4/67H/OdqlaEH/Hk/8A1zX+bVDf63/AldX/wDRYqqg3/HrJfaP/GtcpfSP
WO3Q/Q/yqqva9J+5F9R9R9R6FWrNzyL+5F 7LTbz/j3P8A/64n/9ErKANFmqit1
P1T+ZqRv9dJ/vvwrif6w/7n/n/n/Z
END

     Medical Informatics methods and algorithms in Perl, Python, and Ruby
     </dc:description>
     <dc:date>2010</dc:date>
</rdf:Description>
</rdf:RDF>
```

Figure 18.2   Abbreviated contents of the output file rdf_image.xml

This might be the case when a single RDF file must contain information on multiple different image files. Although it is nice to know that the option of inserting image data into an RDF file is available, in most instances, you will simply point to the external image file (using its Web address), and retrieve the image data from its URL.

18.5 RDF Schema

RDF has a formal way of defining objects (and their properties, but we will not discuss properties here). This is called RDF Schema. You can think of RDF Schema as a dictionary for the terms in an RDF data document. RDF Schema is written in RDF syntax. This means that all RDF Schemas are RDF documents and consist of statements in the form of triples.

The important point about RDF Schemas is that they clarify the relationships among classes of objects in a knowledge domain. Here is an example of Class relationships formally specified as a Schema in RDF:

```
<rdfs:Class rdf:ID="Neoplasm">
<rdfs:subClassOf
rdfs:resource="http://www.w3.org/2000/01/rdf-schema#Class"/>
</rdfs:Class>
```

```
<rdfs:Class rdf:ID="Neural_crest">
<rdfs:subClassOf
neo:resource="#Neoplasm"/>
</rdfs:Class>

<rdfs:Class rdf:ID="Germ_cell">
<rdfs:subClassOf
neo:resource="#Neoplasm"/>
</rdfs:Class>

<rdfs:Class rdf:ID="Mesoderm">
<rdfs:subClassOf
neo:resource="#Neoplasm"/>
</rdfs:Class>

<rdfs:Class rdf:ID="Coelomic">
<rdfs:subClassOf
neo:resource="#Mesoderm"/>
</rdfs:Class>

<rdfs:Class rdf:ID="Sub_coelomic">
<rdfs:subClassOf
neo:resource="#Mesoderm"/>
</rdfs:Class>

<rdfs:Class rdf:ID="Sub_coelomic_gonadal">
<rdfs:subClassOf
neo:resource="#Sub_coelomic"/>
</rdfs:Class>
```

RDF schemas can be transformed into directed graphs (graphs consisting of connected nodes and arcs and directions for the arcs). The process of transforming an RDF Schema into a graphic representation requires a special software application, such as GraphViz.

18.6 Visualizing an RDF Schema with GraphViz

GraphViz is a free, open source application that produces graphic representations of hierarchical structures that are described in the GraphViz scripting language.

As an example, here is the hierarchical organization of the Neoplasm Classification, described in the GraphViz scripting language:

```
digraph G {
size="10,16";
ranksep="1.75";
```

```
node [style=filled color=gray65];
Neoplasm [label="Neoplasm"];
node [style=filled color=lightgray];
EndodermEctoderm
[label="Endoderm\/\nEctoderm"];
NeuralCrest [label="Neural Crest"];
GermCell [label="Germ cell"];
Neoplasm -> EndodermEctoderm;
Neoplasm -> Mesoderm;
Neoplasm -> GermCell;
Neoplasm -> Trophectoderm;
Neoplasm -> Neuroectoderm;
Neoplasm -> NeuralCrest;
node [style=filled color=gray95];
Trophectoderm -> Molar;
Trophectoderm -> Trophoblast;
EndodermEctoderm -> Odontogenic;
EndodermEctodermPrimitive
[label="Endoderm\/Ectoderm\nPrimitive"];
EndodermEctoderm -> EndodermEctodermPrimitive;
Endocrine
[label="Endoderm/Ectoderm\nEndocrine"];
EndodermEctoderm -> Endocrine;
EndodermEctoderm -> Parenchymal;
Odontogenic
[label="Endoderm/Ectoderm\nOdontogenic"];
EndodermEctoderm -> Surface;
MesodermPrimitive
[label="Mesoderm\nPrimitive"];
Mesoderm -> MesodermPrimitive;
Mesoderm -> Subcoelomic;
Mesoderm -> Coelomic;
NeuroectodermPrimitive
[label="Neuroectoderm\nPrimitive"];
NeuroectodermNeuralTube
[label="Central Nervous\nSystem"];
Neuroectoderm -> NeuroectodermPrimitive;
Neuroectoderm -> NeuroectodermNeuralTube;
NeuralCrestMelanocytic
[label="Melanocytic"];
NeuralCrestPrimitive
[label="Neural Crest\nPrimitive"];
```

NeuralCrestEndocrine
[label="Neural Crest\nEndocrine"];
PeripheralNervousSystem
[label="Peripheral\nNervous System"];
NeuralCrestOdontogenic
[label="Neural Crest\nOdontogenic"];
NeuralCrest -> NeuralCrestPrimitive;
NeuralCrest -> PeripheralNervousSystem;
NeuralCrest -> NeuralCrestEndocrine;
NeuralCrest -> NeuralCrestMelanocytic;
NeuralCrest -> NeuralCrestOdontogenic;
GermCell -> Differentiated;
GermCell -> Primordial;
}

By eliminating the lowest level of subclasses, we can generate a simpler schematic (Figure 18.3).

18.7 Obtaining GraphViz

GraphViz is free software. The GraphViz download site is

http://www.graphviz.org/Download.php

Windows® users can download graphviz-2.14.1.exe (5,614,329 bytes). You can install the software by running the .exe file.

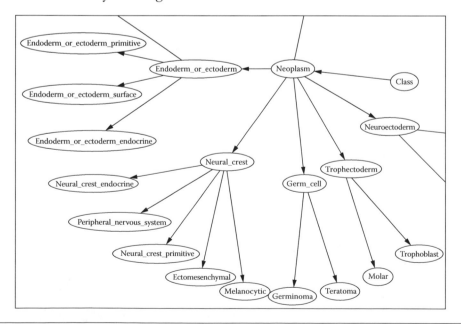

Figure 18.3   A truncated version of the digraph schema.

GraphViz has subapplications: dot, fdp, twopi, neato, and circo. The twopi application, which we use here, creates graphs that have a radial layout.

Extensive information on GraphViz is available at

http://www.graphviz.org/

18.8 Converting a Data Structure to GraphViz

If you work with RDF (and every biomedical professional should understand how RDF is used to specify data), you will want a method that can instantaneously render a schematic of your RDF Schema (ontology) or of any descendant section of your Schema.

Because the GraphViz language is designed with a similar purpose as RDF Schema—to describe the relationships among hierarchical classes of object—it is always possible to directly translate an RDF Schema into the GraphViz language. This is a type of poor man's metaprogramming (using a programming language to generate another program). When an RDF Schema has been translated into the GraphViz language, the GraphViz software can display the class structure as a graph.

18.8.1 Script Algorithm

1. Open the file containing the Schema relationships, in RDF syntax (available at www.julesberman.info/book/schema.txt).
2. Open an output file, to write the transformed class relationships, in the GraphViz language.
3. Print the first lines of the GraphViz file, which begins with a statement indicating that a digraph will follow (instructions for a directed graph), its size, and the length of the separator lines between classes.
4. Parse through the RDF classes, using the end tag "</rdfs:Class>" to indicate the end of one class definition and the beginning of the next class definition.
5. Obtain the name of the class and the name of the class to which the class is a subclass (i.e., the name of the child class's father).

   All of the Schema class statements will have a form equivalent to the following example:

   <rdfs:Class rdf:ID="Neoplasm">
   <rdfs:subClassOf
   rdfs:resource="http://www.w3.org/2000/01/rdf-schema#Class"/>
   </rdfs:Class>

   The class name appears in quotes, after "rdf:ID=". The superclass name appears at the end of a resource statement: "resource="http://www.w3.org/2000/01/rdf-schema#". Use regular expressions to obtain the name of the child class and the father class from each RDF Schema statement.

6. Print to the output file each encountered child, and father class in a GraphViz statement of the following general form:

> father class -> child class;

7. After the schema is parsed, print "}" to the output file, to close the GraphViz script.

Perl Script

```perl
#!/usr/local/bin/perl
open (TEXT, "schema.txt");
open (OUT, ">schema.dot");
$/ = "\<\/rdfs\:Class>";
print OUT "digraph G \{\n";
print OUT  "size\=\"15\,15\"\;\n";
print OUT "ranksep\=\"2\.00\"\;\n";
$line = " ";
while ($line ne "")
  {
  $line = <TEXT>;
  last if ($line !~ /\<rdfs\:/);
  if ($line =~ /\:resource\=\"[a-z0-9\:\/\_\.\-]*\#([a-z\_]+)\"/i)
     {
     $father = $1;
     }
  if ($line =~ /rdf\:ID\=\"([a-z\_]+)\"/i)
     {
     $child = $1;
     }
   print OUT "$father \-\> $child\;\n";
   print "$father \-\> $child\;\n";
   }
print OUT "\}";
exit;
```

Python Script

```python
#!/usr/local/bin/python
import re, string
in_file = open('schema.txt', "r")
out_file = open("schema.dot", "w")
print>>out_file, "digraph G {"
print>>out_file, "size=\"15,15\";"
print>>out_file, "ranksep=\"3.00\";"
clump = ""
for line in in_file:
  namematch = re.match(r'\<\/rdfs\:Class>', line)
  if (namematch):
    father = ""
    child = ""
```

```
    clump = re.sub(r'\n', ' ', clump)
    fathermatch = re.search(r'\:resource\=\"[a-zA-Z0-9\:\/\_\.\-]*
\#([a-zA-Z\_]+)\"', clump)
      if fathermatch:
        father = fathermatch.group(1)
      childmatch = re.search(r'rdf\:ID\=\"([a-zA-Z\_]+)\"', clump)
      if childmatch:
        child = childmatch.group(1)
      print>>out_file, father + " -> " + child + ";"
      clump = ""
    else:
      clump = clump + line
print>>out_file, "}"
exit
```

Ruby Script

```
#!/usr/local/bin/ruby
in_file = File.open("schema.txt", "r")
out_file = File.open("schema.dot", "w")
$/ = "\<\/rdfs\:Class>"
out_file.puts "digraph G \{\n"
out_file.puts "size\=\"15\,15\"\;"
out_file.puts "ranksep\=\"3\.00\"\;"
in_file.each_line do
  |line|
  if line =~ /\:resource\=\"[a-z0-9\:\/\_\.\-]*\#([a-z\_]+)\"/i
    father = $1
  end
  if line =~ /rdf\:ID\=\"([a-z\_]+)\"/i
    child = $1
  end
  if child
    if father
      out_file.puts father + "\-\> " + child + "\;"
    end
  end
end
out_file.puts "\}"
exit
```

18.8.2 Analysis

The output is the script, schema.dot, which is equivalent to the digraph (GraphViz language) script shown at the beginning of this section.

After installing GraphViz, we can create the image schema.png, from the schema.dot specification by invoking the twopi subapplication on a command line.

c:\ftp>twopi -Tpng schema.dot.dot -o schema.png

GraphViz produced the graph shown in Figure 18.3, from a GraphViz script produced by transformations on a RDF Schema.

Exercises

1. Using Perl, Python, or Ruby, write a short script that extracts the Dublin Core RDF from the image header prepared in this chapter, and neatly prints out the RDF tags (metadata) and data.
2. Using Perl, Python, or Ruby, write a script that parses through the image files in a directory, extracting any RDF text included in the headers, and produces a summary RDF document, in proper RDF syntax, containing the Dublin Core RDF data for every image (that contains RDF data).
3. Using Perl, Python, or Ruby, write a script that will parse through the image files in a directory, extracting the Dublin Core RDF (if there is any), and list the filenames of every image file that shares the same creator, and also list the name of the creator.
4. Using any image file, write a script in Perl, Python, or Ruby that creates a simple XML file containing the entire binary image file encoded as Base64 text as the contents for a single tag in the XML file, and the entire header of the file as the contents of another tag in the file.
5. Write another script that will take the XML file from Exercise 4, extract the Base64 text from the file, convert the text to binary format, and display the image in a viewer.

PART IV
MEDICAL DISCOVERY

19

CASE STUDY

Emphysema Rates

Much of computational medicine can be described as a solution in search of a question. We have many ways of analyzing data, but we often lack important questions that can be solved with our available data. Let us focus on the kinds of biological questions that can be approached with the CDC data.

Alpha-1 antitrypsin disease is a prototypical serpinase disease (disease due to deficiencies or abnormalities in the synthesis of serine proteinases). People with this disorder are homozygous for mutations in the alpha-1 antitrypsin gene. The full-blown disease is characterized by cirrhosis and emphysema. The pathogenesis of this disease is somewhat complex, because there are a variety of different possible mutations of the gene, and the clinical manifestations vary somewhat with the mutation type. The cirrhosis is apparently due to the intracellular accumulation of abnormal alpha-1 antitrypsin molecules within hepatocytes, and the emphysema is apparently the result of destructive effects of inflammation-inducing intrapulmonary trypsin levels, unopposed by antitrypsin.

As is the case in most rare recessive genetic disorders, heterozygous mutations in the alpha-1 antitrypsin gene are found as common gene variants in the general population.

If a double-dose (homozygous) of an altered gene causes disease, what is the effect of a single (heterozygous) gene variant? Gene variations may be responsible for differences in the pathogenesis of disease among members of the apparently healthy public. About 15% of smokers develop COPD (chronic obstructive pulmonary disease) or emphysema. Why does one smoker develop COPD, while another smoker escapes pulmonary toxicity? Might the difference be accounted for by gene variations, and might a key gene be the alpha-1 antitrypsin gene?

Several researchers have provided data indicating that heterozygous carriers of alpha-1 antitrypsin mutations are at increased risk for developing emphysema.*

---

* Lieberman, J. Heterozygous and homozygous alpha-1-antitrypsin deficiency in patients with pulmonary emphysema. New Eng J Med 281:279–284, 1969; Stevens, P.M., Hnilica, V., Johnson, P.C., Bell, R.L. Pathophysiology of hereditary emphysema. Ann Intern Med 74:672-680, 1971.

Population studies indicate that the African-American population has much lower levels of alpha-1 antitrypsin disease gene variants than whites, the most prevalent mutations occurring in people with European ancestry.*

We hypothesize that if alpha-1 antitrypsin disease mutations play a significant contributory role in the pathogenesis of emphysema in the general population, we can expect to see fewer emphysema cases in African-Americans (who are unlikely to be heterozygous for alpha-1 antitrypsin diseases mutations) than in the white population. We can test this hypothesis by determining the percentage of African-Americans who die in the United States of emphysema, and comparing that number with the percentage of white Americans who die of the disease.

19.1 Script Algorithm

1. Open the CDC mortality data for 1999 (file: Mort99us.dat; download instructions provided in Chapter 13)
2. Parse through each record (i.e., through each line in the file), keeping a tally of the total number of records (by incrementing a variable at each parsing loop).
3. From each line, extract the substring that contains the disease codes for the death record (bytes 160 to 300).
4. The ICD codes for emphysema and COPD (chronic obstructive pulmonary disease) begin with "J4," followed by "3" or "4." Determine whether the substring containing the ICD codes matches the codes for emphysema or COPD.
5. Race is assigned a two-digit code, 01 for White and 02 for Black, at bytes 60 and 61 of each record. Extract the substring containing the race code. If the code corresponds to White ("01"), increment the variable containing the running tally of the encountered death records of white persons by 1. If the code corresponds to Black ("02"), increment the variable containing the running tally of the encountered death records of black persons by 1.
6. If the record contains a diagnosis of emphysema or COPD, and the record pertains to a white person (i.e., has a code of "01" at bytes 60 and 61), increment the running tally of white persons with emphysema or COPD by 1. If the record contains a diagnosis of emphysema or COPD, and the record pertains to a black person (i.e., has a code of "02" at bytes 60 and 61), increment the running tally of black persons with emphysema or COPD by 1.
7. After the entire mortality file has been parsed, you will have collected the following variables: total number of records, total number of records of white persons, total number of records of black persons, total number of black persons

---

* DeCroo, S., Kamboh, M.I., Ferrell, R.E. Population genetics of alpha-1-antitrypsin polymorphism in US whites, US blacks and African blacks. Hum Hered 41:215–221, 1991; Hutchison, D.C.S. Alpha-1-antitrypsin deficiency in Europe: Geographical distribution of Pi types S and Z. Resp Med 92:367–377, 1998.

with emphysema or COPD, and total number of white persons with emphysema or COPD. Use these variables to determine the fraction of blacks with emphysema or COPD and the fraction of whites with emphysema or COPD.

Perl Script

```perl
#!/usr/local/bin/perl
open (ICD, "Mort99us.dat");
$line = " ";
while ($line ne "")
  {
  $line = <ICD>;
  $count++;
  $codesection = substr($line,161,140);
  $race = substr($line,59,2);
  $whitecount++ if ($race eq "01");
  $blackcount++ if ($race eq "02");
  if ($codesection =~ /J43/)
    {
    $whiteemp++ if ($race eq "01");
    $blackemp++ if ($race eq "02");
    }
  }
close ICD;
$whiteempfrac = 100 * ($whiteemp / $whitecount);
$blackempfrac = 100 * ($blackemp / $blackcount);
print "Total records in file is $count\n";
print "Total African-Americans in file is $blackcount\n";
print "Total Whites in file is $whitecount\n";
print "Total African-Americans with emphysema is $blackemp\n";
print "Total Whites with emphysema is $whiteemp\n";
print "Percent African-Americans with emphysema is ";
print substr($blackempfrac,0,4) . "\n";
print "Percent Whites with emphysema is ";
print substr($whiteempfrac,0,4) . "\n";
exit;
```

Python Script

```python
#!/usr/local/bin/python
import re, string
in_file = open("c:\\big\\Mort99us.dat")
count = 0
whitecount = 0
blackcount = 0
whiteemph = 0
blackemph = 0
for line in in_file:
  count = count + 1
  codesection = line[161:301]
```

```
  race = line[59:61]
  if race == "01":
    whitecount = whitecount + 1
  if race == "02":
    blackcount = blackcount + 1
  codematch = re.search(r'J43', codesection)
  if codematch:
    if race == "01":
      whiteemph = whiteemph + 1
    if race == "02":
      blackemph = blackemph + 1
in_file.close()
whiteempfrac = str(100 * (float(whiteemph) / whitecount))
blackempfrac = str(100 * (float(blackemph) / blackcount))
print "Total records in file is " + str(count)
print "Total African-Americans in file is " + str(blackcount)
print "Total Whites in file is " + str(whitecount)
print "Total African-Americans with emphysema is " + str(blackemph)
print "Total Whites with emphysema is " + str(whiteemph)
print "Percent African-Americans with emphysema is " +
blackempfrac[0:4]
print "Percent Whites with emphysema is " + whiteempfrac[0:4]
exit
```

Ruby Script

```ruby
#!/usr/local/bin/ruby
require 'mathn'
in_file = File.open("c:/big/Mort99us.dat")
count = 0
whitecount = 0
blackcount = 0
whiteemph = 0
blackemph = 0
in_file.each_line do
  |line|
  count = count + 1
  codesection = line.slice(161,140)
  race = line.slice(59,2)
  whitecount = whitecount + 1 if race.eql?("01")
  blackcount = blackcount + 1 if race.eql?("02")
  if (codesection =~ /J43/)
    whiteemph = whiteemph + 1 if race.eql?("01")
    blackemph = blackemph + 1 if race.eql?("02")
  end
end
in_file.close
whiteempfrac = (100 * ((whiteemph / whitecount).to_f)).to_s
blackempfrac = (100 * ((blackemph / blackcount).to_f)).to_s
puts "Total records in file is " + count.to_s
```

```
puts "Total African-Americans in file is " + blackcount.to_s
puts "Total Whites in file is " + whitecount.to_s
puts "Total African-Americans with emphysema is " + blackemph.to_s
puts "Total Whites with emphysema is " + whiteemph.to_s
puts "Percent African-Americans with emphysema is " + blackempfrac.
slice(0,4)
puts "Percent Whites with emphysema is " + whiteempfrac.slice(0,4)
exit
```

19.2 Analysis

Here is the output from the script:

Total records in file is 2394871

Total African-Americans in file is 285276

Total Whites in file is 2064169

Total African-Americans with emphysema is 2125

Total Whites with emphysema is 32595

Percent African-Americans with emphysema is 0.74

Percent Whites with emphysema is 1.57

The Perl script examines 2.3 million death records in the CDC data set, and informs us that African-Americans have about half the rate of emphysema and COPD compared with the white population. This observation is consistent with our hypothesis that the alpha-1 antitrypsin gene variant increases the risk of emphysema in the general population.

Does this observation prove our hypothesis? Absolutely not. The same observation could be explained by many different hypotheses. We have shown, with a large number of cases (nearly a quarter million emphysema/COPD cases), that African-Americans have less emphysema than whites. This observation may justify further laboratory work that conclusively determines whether the alpha-1 antitrypsin gene contributes to the development of emphysema in heterozygous carriers.

This is the kind of analysis that uses existing CDC mortality data sets to develop and test a hypothesis. In the next chapter, we will relate our available data with a graphic representation.

Exercises

1. Rewrite the script to include chronic bronchitis along with emphysema. The relevant ICD codes are

    J41 Simple and mucopurulent chronic bronchitis

    J41.0 Simple chronic bronchitis

    J41.1 Mucopurulent chronic bronchitis

    J41.8 Mixed simple and mucopurulent chronic bronchitis

    J42 Unspecified chronic bronchitis

J43 Emphysema

J43.0 MacLeod's syndrome

J43.1 Panlobular emphysema

J43.2 Centrilobular emphysema

J43.8 Other emphysema

J43.9 Emphysema, unspecified

J44 Other chronic obstructive pulmonary disease

J44.0 Chronic obstructive pulmonary disease with acute lower respiratory infection

J44.1 Chronic obstructive pulmonary disease with acute exacerbation, unspecified

J44.8 Other specified chronic obstructive pulmonary disease

J44.9 Chronic obstructive pulmonary disease, unspecified

2. Using Perl, Python, or Ruby, rewrite the script to determine the emphysema rates in Hispanics.

3. Using Perl, Python, or Ruby, rewrite the script to compare the emphysema rates in white men, white women, African-American men, and African-American women.

4. Emphysema is a disease in which the airway spaces (the small spaces where oxygen passes across a thin layer of tissue into the deoxygenated red blood cells) are destroyed. Pathologists often divide emphysema into two morphologic types: centrilobular emphysema (J43.2) and panlobular emphysema (J43.1). In centrilobular emphysema, destruction of the airway spaces occurs primarily in the center of the lobules (near the bronchi). In panlobular emphysema, airway destruction is seen uniformly throughout the lobule. Using Perl, Python, or Ruby, write a script to determine which of the two morphologic types of emphysema is more common in the U.S. population.

5. Modify your script from Exercise 4 to determine that average age at death of each type of emphysema: centrilobular emphysema (J43.2), and panlobular emphysema (J43.1).

6. Modify your script from Exercise 5 to determine the gender differences in the total number of occurrences and the average age at death of the two types of emphysema.

7. Modify your script from Exercise 6 to determine the difference between black persons and white persons in the total number of occurrences and the average age at death of the two types of emphysema.

20

CASE STUDY

Cancer Occurrence Rates

The SEER data sets contain information on the race of each individual record. Because there are over 3.7 million records in the SEER database, we can ask questions about racial differences in the occurrences of every type of tumor, even rare tumors, and rare variants of common tumors.

What is the value of race-based analysis? On occasion, a difference in the rate of occurrence of tumors among races may be due to identifiable (and mutable) exposure to carcinogens, or living conditions. Sometimes, socioeconomic conditions account for the differences. Occasionally, the differences lie in genetic traits that are found more often in one race than in another, and this may lead to an intervention that modifies the effect of the trait on the development of cancer.

All of these discoveries begin with finding a difference between races, and the best way of finding differences is by starting with a data set containing millions of cancer records.

20.1 Script Algorithm

1. SEER cancer diagnoses are coded with the ICD-O (International Classification of Diseases–Oncology), discussed in Chapter 6, Section 6.2. The ICD-O dictionary file is icdo3.txt. Create a dictionary object from the icdo3.txt file, in which the keys are code numbers and the values are the corresponding neoplasm names.
2. Parse through SEER files, line by line. Each line of a SEER file is the record of a cancer occurrence, and there are over 3.7 million lines that will be parsed. Instructions for obtaining the free, public use SEER files is found in the appendix. In this example script, the SEER files are found in the c:\big\ seer2006 subdirectory.
3. As each line of the file is parsed, extract the 5-character substring that begins at character 48 and the 5-character substring that begins at character 53. These represent the ICD-O code for the record. The string beginning at character 53 is the code for the newer version of ICD-O. If this string has a code that is contained in the version of ICD-O that we are using (in the icdo3.txt file), we will use this code, rather than the code contained in the substring that begins at character 48.

RACE/ETHNICITY

NAACCR ITEM #: N/A
SEER*Stat Name: Race/ethnicity
Item Length: 2

Field Description: This field is created from Race1 and the Indian Health Service (IHS) Link
variables from the NAACCR File Format. If Race1 is white and there is a
positive IHS link, then Race/ethnicity is set to American Indian/Alaskan
Native, otherwise Race/Ethnicity is set to the Race1 value.

Codes

01	White
02	Black
03	American Indian, Aleutian, Alaskan Native or Eskimo (includes all indigenous populations of the Western hemisphere)
04	Chinese
05	Japanese
06	Filipino
07	Hawaiian
08	Korean (Effective with 1/1/1998 dx)
09	Asian Indian, Pakistani (Effective with 1/1/1988 dx)
10	Vietnamese (Effective with 1/1/1988 dx)
11	Laotian (Effective with 1/1/1988 dx)
12	Hmong (Effective with 1/1/1988 dx)
13	Kampuchean (including Khmer and Cambodian) (Effective with 1/1/1988 dx)

Figure 20.1   Race codes for the SEER cancer files, SEERDIC6.PDF.

4. Create a dictionary object whose keys are the encountered neoplasm codes, and whose values are the incremented tally of the number of encountered SEER records that contain the code.

5. Extract, from each record, the two-character sequence of characters 20 and 21. This two-character sequence contains the "race" field of the SEER record. The assignments for the race sequence are shown in Figure 20.1. The code for "Black" is "02", and the code for "White" is "01".

6. If the code is "02", increment the variable that contains a running tally of number of black persons with cancer by 1. If the code is "02", increment by 1 the dictionary object that has as values a running tally of the number of black persons that have a particular cancer and as keys the code for the particular cancer.

7. If the code is "01", increment, by 1 the variable that contains a running tally of the number of white persons with cancer. If the code is "01", increment by 1 the dictionary object that has as values a running tally of the number of white

persons that have a particular cancer and as keys the code for the particular cancer.

8. After all the SEER files have been parsed, we are left with a variable with the total number of black persons with ICD-O encoded cancers; a variable with the total number of white persons with ICD-O encoded cancers; a dictionary object with ICD-O codes encountered in the SEER files as the keys and the total number of occurrences of the cancer in black persons as the values; and a dictionary object with ICD-O codes encountered in the SEER files as the keys and the total number of occurrences of the cancer in white persons as the values.

9. Print to the monitor the total number of black persons with cancer in the SEER data files. Print to the monitor the total number of white persons with cancer in the SEER data files.

10. Iterate over the key–value pairs in the dictionary object containing the neoplasm codes and occurrence numbers of the neoplasms occurring in black persons.

    For those neoplasms that have at least 5 occurrences in black persons and 5 occurrences in white persons, determine the relative frequency of occurrence of neoplasms found in the white population compared with the black population. This proportion is obtained using the following equation:

    > relative frequency of occurrence = (number of tumors occurring in whites divided by the number of white persons) divided by (number of tumors occurring in black persons divided by the number of black persons).

    For example, if the relative frequency of a neoplasm were "10", then the neoplasms occurred 10 times more frequently in the white population than in the black population in the SEER data files. If the tumor accounted for the same fraction of total cancer cases in the white and black populations, it would have a ratio of 1.

11. Open an external file. For each type of neoplasm, print to the external file a formatted line containing the relative frequency of occurrence of the neoplasm, the number of occurrences of the neoplasm in the SEER data files, and the name of the neoplasm.

Perl Script

```perl
#!/usr/local/bin/perl
open (ICD, "c\:\\ftp\\icdo3\.txt");
$line = " ";
$black_count = 0;
$white_count = 0;
while ($line ne "")
  {
```

```perl
$line = <ICD>;
if ($line =~ /([0-9]{4})\/([0-9]{1}) +/o)
   {
   $code = $1 . $2;
   $term = $';
   $term =~ s/ *\n//o;
   $dictionary{$code} = $term;
   }
}
close ICD;
opendir(FTPDIR, "c\:\\big\\seer2006") || die ("Unable to open
directory");
@files = readdir(FTPDIR);
closedir(FTPDIR);
chdir("c\:\\big\\seer2006");
foreach $datafile (@files)
  {
  open (TEXT, $datafile);
  $line = " ";
  while ($line ne "")
     {
     $line = <TEXT>;
     $code = substr($line, 47, 5);
     $code2 = substr($line, 52, 5);
     if (exists($dictionary{$code2}))
        {
        $code = $code2;
        }
     if (exists($dictionary{$code}))
        {
        $code_count{$code}++;
        if (substr($line,19,2) eq "02")    #02 means black
          {
          $black_count++;
          $bl_count{$code}++;
          }
        if (substr($line,19,2) eq "01")    #01 means white
          {
          $white_count++;
          $wh_count{$code}++;
          }
        }
     }
  close(TEXT);
  }
print "\nNumber of black persons with cancer is $black_count\n";
print "Number of white persons with cancer is $white_count\n";
```

```perl
open(OUT, ">c\:\\ftp\\seer_out.txt");
while((my $key, my $value) = each (%bl_count))
    {
    if ($bl_count{$key} > 5)
        {
        if ($wh_count{$key} > 5)
            {
            $whfract = ($wh_count{$key} / $white_count) / ($bl_
count{$key} / $black_count);
            printf OUT ("%05.2f %-6.6d %-55.55s\n", $whfract,
$code_count{$key}, $dictionary{$key});
            }
        }
    }
exit;
```

Python Script

```python
#!/usr/local/bin/python
import sys, os, re, string
icd_in = open("c:/ftp/icdo3.txt", "r")
dictionary = {}
bl_count = {}
wh_count = {}
code_count = {}
black_count = 0
white_count = 0
for line in icd_in:
    code_has = re.search(r'([0-9]{4})\/([0-9]{1}) +(.+)', line)
    if code_has:
        code = code_has.group(1) + code_has.group(2)
        term = code_has.group(3)
        term = string.rstrip(term)
        dictionary[code] = term
        bl_count[code] = 0
        wh_count[code] = 0
        code_count[code] = 0
icd_in.close
filelist = os.listdir("c:/big/seer2006")
os.chdir("c:/big/seer2006")
for file in filelist:
    infile = open(file,'r')
    for line in infile:
        code = line[47:52]
        code2 = line[52:57]
        if dictionary.has_key(code2):
            code = code2
        if dictionary.has_key(code):
```

```
      code_count[code] = code_count[code] + 1
      if (line[19:21] == "02"):                    #02 means black
         black_count = black_count + 1
         bl_count[code] = bl_count[code] + 1
      if (line[19:21] == "01"):
         white_count = white_count + 1
         wh_count[code] = wh_count[code] + 1
   infile.close()
os.chdir("c:/ftp/py")
print "Number of black persons with cancer is " + str(black_count)
print "Number of white persons with cancer is " + str(white_count)
print
seer_out = open("c:/ftp/seer_out.txt", "w")
for key,value in bl_count.iteritems():
   if (bl_count[key] > 5):
      if (wh_count[key] > 5):
         wbfract = (float(wh_count[key]) / float(white_count)) /
(float(bl_count[key]) / float(black_count))
         print>>seer_out, "%05.2f %-6.6d %-55.55s" % (wbfract,
code_count[key], dictionary[key])
exit
```

Ruby Script

```ruby
#!/usr/local/bin/ruby
require 'mathn'
icd_in = File.open("c:/ftp/icdo3.txt", "r")
dictionary = {}
bl_count = Hash.new(0)
wh_count = Hash.new(0)
code_count = Hash.new(0)
white_count = 0
black_count = 0
icd_in.each_line do
   |line|
   if line =~ /([0-9]{4})\/([0-9]{1}) +/
      code = $1 + $2
      term = $'
      term = term.sub(/ *\n/, "") if term =~ / *\n/
      dictionary[code] = term
   end
end
icd_in.close
seer_out = File.open("c:/ftp/seer_out.txt", "w")
filelist = Dir.glob("c:/big/seer2006/*.TXT")
filelist.each do
 |filepathname|
 seer_in = File.open(filepathname)
 seer_in.each_line do
   |line|
```

```
    code = line.slice(47, 5)
    code2 = line.slice(52, 5)
    if dictionary.has_key?(code2)
      code = code2
    end
    if dictionary.has_key?(code)
      code_count[code] = code_count[code] + 1
      if line.slice(19,2).eql?("02")  #02 means black race
        black_count = black_count + 1
        bl_count[code] = bl_count[code] + 1
      end
      if line.slice(19,2).eql?("01")  #01 means white race
        white_count = white_count + 1
        wh_count[code] = wh_count[code] + 1
      end
    end
  end
  seer_in.close
end
puts "Number of black persons with cancer is " + black_count.to_s
puts "Number of white persons with cancer is " + white_count.to_s
puts
bl_count.each_pair do
  |key, value|
  if bl_count[key] > 5
    if wh_count[key] > 5
      hbfract = (wh_count[key].to_f / white_count.to_f) /
(bl_count[key].to_f / black_count.to_f)
      seer_out.printf "%05.2f %-6.6d %-55.55s\n", hbfract,
code_count[key], dictionary[key]
    end
  end
end
exit
```

20.2 Analysis

A partial output, which I have sorted by relative frequencies, is shown in Figure 20.2.

The first column is the case occurrence ratio (white/black). The second column is the number of cases, for each tumor type, in the SEER public use data sets. The fourth column is the ICD-O (International Classification of Diseases–Oncology) neoplasm term.

The figure shows the first 12 and the last 16 entries from the output file, which contains a list of hundreds of tumors. These selected entries cover the most extreme differences between black persons and white persons in the rate of occurrence of tumors. A file of the sorted raw data is available for download at

http://www.julesberman.info/book/seerblwh.txt

```
00.11    000044    Pigmented dermato fibrosarcoma protuberans
00.16    000022    Malignant placental site trophoblastic tumor
00.17    000137    Adult T-cell leukemia/lymphoma (HTLV-1 pos.)
00.19    000017    Primary effusion lymphoma
00.19    000258    Adenoma, NOS
00.22    000024    Hepatosplenic gamma-delta cell lymphoma
00.23    000057    Granular cell tumor, malignant
00.24    000037    Adamantinomatous craniopharyngioma
00.26    000078    Collecting duct carcinoma
00.26    000068    Thymoma, type AB, malignant
00.26    000093    Ameloblastoma, malignant
00.27    000248    Cyst-associated renal cell carcinoma
         .

LOTS MORE HERE

         .
05.33    001140    Ewing sarcoma
05.66    003242    Embryonal carcinoma, NOS
05.94    000422    Epithelioid cell melanoma
05.96    000569    Choriocarcinoma combined w/ other germ cell
                   elements
06.09    000630    Seminoma, anaplastic
08.56    001873    Merkel cell carcinoma
08.60    000860    Paget disease, extramammary
11.55    000699    Mixed epithel. & spindle cell melanoma
13.07    051617    Malignant melanoma, NOS
16.75    001357    Spindle cell melanoma, NOS
25.81    009096    Nodular melanoma
30.62    007741    Lentigo maligna melanoma
32.93    024502    Melanoma in situ
39.46    042918    Superficial spreading melanoma
40.83    019950    Lentigo maligna
77.01    004805    Superficial spreading melanoma, in situ
```

Figure 20.2 Relative frequencies of occurrence of neoplasms in black persons and white persons; only a partial output is shown.

Let us look at the last 9 entries from the output file:

White/black ratio, No. of cases, ICD-O Diagnosis
11.55 000699 Mixed epithel. & spindle cell melanoma
13.07 051617 Malignant melanoma, NOS
16.75 001357 Spindle cell melanoma, NOS
25.81 009096 Nodular melanoma
30.62 007741 Lentigo maligna melanoma
32.93 024502 Melanoma in situ
39.46 042918 Superficial spreading melanoma
40.83 019950 Lentigo maligna
77.01 004805 Superficial spreading melanoma, in situ

The rate of occurrence of in situ superficial spreading melanoma is 77 times higher in white persons than in black persons. Of the top nine tumors that occur disproportionately more often in white persons than in black persons, all nine are types of melanoma. Of course, this finding is not unexpected; melanin protects skin from the short-term and long-term harmful effects of ultraviolet light: sunburn, solar elastosis, epidermal skin cancer (primarily squamous cell carcinoma and basal cell carcinoma), and melanoma. It has long been known that black persons have a lower risk of melanoma than white persons, and that fair-skinned white persons (particularly Irish, and red-haired white persons) and anyone who sunburns easily has a higher risk of developing melanoma than white persons with black hair or a somewhat darker complexion.

The value of comparing melanoma incidence in black persons and white persons in the SEER data set comes from the large numbers of occurrences and the variant tumor types collected. We see remarkable internal consistency in the list. Where a tumor appears, it is often closely followed by a variant of the same tumor. This indicates that closely related tumors, which have the same general cell type (in this case, tumors of melanocyte origin), most likely have the same biological causes. Otherwise, why would they aggregate in the list?

One type of melanoma occurs in white persons and black persons at about the same occurrence rate.

White/black ratio, No. of cases, ICD-O Diagnosis
00.90 000940 Acral lentiginous melanoma, malig.

Acral lentiginous melanoma is a variant of melanoma that occurs in nonpigmented skin: the sole of the foot, the palm of the hand, under fingernails or toenails. Because black persons and white persons have similar exposure to ultraviolet light on acral skin, you would expect that black persons and white persons would have similar occurrence rates of acral lentiginous melanomas. This is precisely the case. The SEER data set confers biologic consistency to the hypothesis that the differences in the occurrence rates of melanoma in white persons and black persons is based on the protective effect of melanin in black persons. Moreover, the SEER data set includes a total of 940 cases of acral lentiginous melanoma, divided almost equally among black persons or white persons. This is a very large number of cases of a relatively rare tumor. The large number of cases adds credibility to the biological conclusions.

The top cases in the output are the tumors that occur disproportionately more often in black persons than in white persons. Leading the list is a curious and rare tumor, Pigmented dermatofibrosarcoma protuberans

White/black ratio, No. of cases, ICD-O Diagnosis
00.11 000044 Pigmented dermatofibrosarcoma protuberans

This tumor occurs 10 times more frequently in black persons than in white persons. What might account for this difference? A pigmented dermatofibrosarcoma protuberans is a variant of dermatofibrosarcoma protuberans that happens to contain some melanin. Because the cell of origin of dermatofibrosarcoma is not a melanocyte, the origin of the melanin could certainly be of nonneoplastic origin (e.g., the result of melanin produced in the overlying skin and dropping into the tumor matrix). The high rate of occurrence of pigmented melanoma in black persons, compared with white persons, strengthens the hypothesis that the melanin in the tumor is a secondary phenomenon. Persons with a lot of skin melanin would be more likely to deposit melanin within an underlying skin tumor than would persons with very little skin melanin.

Decades of histologic analysis of these tumors provided very little insight into the source of melanin in these tumors. The discovery, from the SEER data, that black persons have 10 times the occurrence rate of pigmented dermatofibrosarcoma protuberans than white persons indicates that the source of pigmentation in this tumor is the overlying skin.

Another interesting finding in the data relates to Ewing sarcoma. Ewing sarcoma is a rare, malignant tumor that occurs in children and young adults, often arising in bones. Generations of pathologists have been taught that Ewing's sarcoma almost never occurs in black persons. Here are our findings:

White/black ratio, No. of cases, ICD-O Diagnosis
05.33 001140 Ewing sarcoma

The SEER data indicate that among 1,140 cases, Ewing sarcoma occurs 5.33 times more commonly in white persons than in black persons. The data does not support the assertion that Ewing tumor never (or rarely) occurs in black persons.

Exercises

1. Using Perl, Python, or Ruby, modify the script from this chapter to display the average age of occurrence in the black population and in the white population, for each neoplasm.
   Hint:
   Age in Perl: $age_at_dx = substr($line,24,3);
   Age in Python: age_at_dx = int(line[24:27])
   Age in Ruby: age_at_dx = line.slice(24,3)
2. Using Perl, Python, or Ruby, modify the script from this chapter to display the average age of occurrence of each neoplasm, and the total number of occurrences of the neoplasms, in black males and black females.
3. Modify the script to produce a comparison between black persons and Chinese persons contained in the SEER data files.
   Hint: Substitute Chinese for White, and "04" for "01" (in the race substring).
4. Modify the script to produce an output in reverse sort (i.e., largest number to smallest number) on the relative frequencies of occurrence of tumors.

21

CASE STUDY

Germ Cell Tumor Rates across Ethnicities

In Chapter 20, Section 20.1, we compared the occurrences of all types of tumors in black persons and white persons. Using the output file, seerblwh.txt, we can see that two germ cell tumors (teratocarcinoma and seminoma) occurred about five times more frequently in white persons than in black persons:

RATIO	NUMBER OF OCCURRENCES	TUMOR NAME
05.03	001732	Teratocarcinoma
04.35	010298	Seminoma, NOS

I wondered whether the same disproportionate occurrence of germ cell tumors is seen in white Hispanic persons compared with black persons?

I went to the SEER site and used SEER's public query engine to see if this observation could be verified.

The SEER query site is at

http://seer.cancer.gov/canques/index.html

The SEER search engine supports queries over a small number of parameters. I searched for tumors in males, in testes, comparing white Hispanics with black persons. A simple interface permits these selections.

The SEER interface produced a list of input parameters (Figure 21.1). The same interface produced a bar chart of results (Figure 21.2).

You may be wondering, if I am interested in germ cell tumors, why did I do a query on tumors of the testes? I did this because the SEER interface does not allow me to query specific types of germ cell tumors or any specific testicular tumor. I know that the vast majority of testicular tumors are germ cell tumors, so I guessed that if there were a difference in the incidence of testicular tumors in white Hispanics compared with black persons, the difference would be due to a difference in the rate of occurrence of germ cell tumors at this site.

And that is what happened. The SEER output demonstrated that white Hispanics had a much higher incidence of testicular tumors, compared with African-Americans.

With a little work, we can write our own script that goes much further than the public SEER database. We can determine the ratio of occurrence for each specific

· **SEER Incidence - Crude Rates for Additional Races and 13 Registries, 1992–2005**
Selections:
    Statistic type = Crude rate;
    SEER registry = Total (registries depend on race/ethnicity);
    Site = Testis;
    Year of diagnosis = 1992–2005;
    Sex = Male;
    Age at diagnosis = All ages;

Figure 21.1   Query parameters for the SEER search engine, specifying testicular tumors occurring between 1992 and 2005, among males of any age.

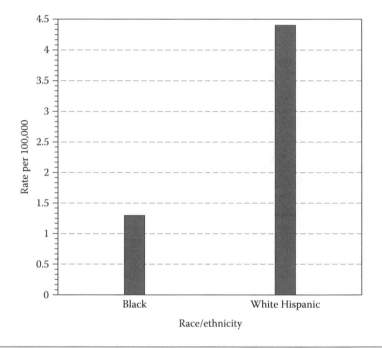

Figure 21.2   SEER search engine bar chart output, indicating the rates of testicular cancer in men, comparing black persons and white Hispanic persons.

germ cell tumor, in black persons and Hispanic persons, for any anatomic location, or for all anatomic locations.

21.1 Script Algorithm

1. Use the icdo3.txt file, containing the list of neoplasm names and codes used in the SEER data files, to create a dictionary object in which the keys are code numbers and the values are the corresponding neoplasm names.
2. Parse through your downloaded SEER files, line by line (over 3.7 million lines). In this example script, the SEER files are found in the \seer2006 subdirectory.

3. As each line of the file is parsed, extract the 5-character substring that begins at character 48 and the 5-character substring that begins at character 53. These represent the ICD-O code for the record. The string beginning at character 53 is the code for the newer version of ICD-O. If this string has a code that is contained in the version of ICD-O that we are using (in the icdo3.txt file), we will use this code rather than the code contained in the substring that begins at character 48.

4. Determine whether the record code corresponds to one of the codes of the germ cell tumors.

   Here is the section of the ICD-O listing for the germ cell tumors:

   ! GERM CELL TUMORS 906 9060/3 Dysgerminoma

   9061/3 Seminoma, NOS

   9062/3 Seminoma, anaplastic

   9063/3 Spermatocytic seminoma

   9064/3 Germinoma

   9065/3 Germ cell tumor, nonseminomatous

   EMBRYONAL CARCINOMA, NOS 907 9070/3 Embryonal carcinoma, NOS

   9071/3 Yolk sac tumor

   9072/3 Polycmbryoma

   ! TERATOMA 908 9080/3 Teratoma, malignant, NOS

   9081/3 Teratocarcinoma

   9082/3 Malignant teratoma, undiff.

   9083/3 Malignant teratoma, intermediate

   9084/3 Teratoma with malig. transformation

   9085/3 Mixed germ cell tumor

All of the germ cell tumors have 5-digit codes that begin with 906, 907, or 908. With a regular expression, match every record code and exclude records that do not contain a neoplasm code beginning with "90" followed by "6", "7", or "8".

5. SEER records contain a byte field (character 22) that specifies Hispanic status (Figure 21.3). Non-Hispanics are designated as "0", and Hispanics are designated with digits 1–9.

6. SEER records contain a two-character sequence of characters 20 and 21, the "race" field of the seer record. The code for "Black" is "02", and the code for "White" is "01".

7. As each record is parsed, determine whether the record is of a white Hispanic person (i.e., Hispanic code of nonzero and a race code of "01") or a black non-Hispanic person.

Codes	
0	Non-Spanish/Non-Hispanic
1*	Mexican (includes Chicano)
2*	Puerto Rican
3*	Cuban
4*	South or Central American (except Brazil)
5*	Other specified Spanish/Hispanic origin (includes European; excludes Dominican Republic)
6	Spanish, NOS; Hispanic, NOS; Latino, NOS *(There is evidence, other than surname or maiden name, that the person is Hispanic but he/she cannot be assigned to any of the categories 1–5).*
7**	Spanish surname only (effective with diagnosis on or after 1/1/1994) *(The only evidence of the person's Hispanic origin is the surname or maiden name and there is no contrary evidence that the patient is not Hispanic.)*
8	Dominican Republic (effective with diagnosis on or after 1/1/2005)
9	Unknown whether Spanish/Hispanic or not

Figure 21.3   SEER Hispanic ethnicity codes.

8. Create a dictionary object whose keys are the encountered neoplasm codes, and whose values are the incremented tally of the number of encountered SEER records that contain the code.

9. Create another dictionary object whose keys are the codes of germ cell tumors and whose values are the running tally of black non-Hispanic persons with the diagnosis.

10. Create another dictionary object whose keys are the codes of germ cell tumors and whose values are the running tally of white Hispanic persons with the diagnosis.

11. After all of the SEER files have been parsed, we are left with a variable containing the total number of black non-Hispanic persons with germ cell tumors; a variable containing the total number of white Hispanic persons with germ cell tumors; a dictionary object with ICD-O codes for the germ cell tumors as keys, and with the total number of occurrences of the cancer in black non-Hispanic persons as the values; a dictionary object with ICD-O codes for the germ cell tumors as keys, and the total number of occurrences of the cancer in white Hispanic persons as the values.

12. Print to the monitor the total number of black persons with cancer in the SEER data files. Print to the monitor the total number of white Hispanic persons with cancer in the SEER data files.

13. Iterate over the key–value pairs in the dictionary object containing the neoplasm codes and occurrence numbers of the neoplasms occurring in black persons.

    For those neoplasms that occur at least 5 times in black non-Hispanic persons and 5 times in white Hispanic persons, determine the relative frequency of occurrence of neoplasms found in the white Hispanic population compared with the black non-Hispanic population. This proportion is obtained using the following equation:

> relative frequency of occurrence = (number of tumors occurring in white Hispanics divided by the number of white Hispanic persons) divided by (number of tumors occurring in black non-Hispanic persons divided by the number of black non-Hispanic persons).

For example, if the relative frequency of a neoplasm were "10", then the neoplasms occurred 10 times more frequently in the white Hispanic population than in the black non-Hispanic population in the SEER data files.

14. Open an external file. For each type of neoplasm, print to the external file a formatted line containing the relative frequency of occurrence of the neoplasm, the number of occurrences of the neoplasm in the SEER data files, and the name of the neoplasm.

Perl Script

```
#!/usr/local/bin/perl
open (ICD, "c\:\\ftp\\icdo3\.txt");
$line = " ";
$black_count = 0;
$hisp_count = 0;
while ($line ne "")
  {
  $line = <ICD>;
  if ($line =~ /([0-9]{4})\/([0-9]{1}) +/o)
    {
    $code = $1 . $2;
    $term = $';
    $term =~ s/ *\n//o;
    $dictionary{$code} = $term;
    }
  }
close ICD;
opendir(FTPDIR, "c\:\\big\\seer2006") || die ("Unable to open
directory");
@files = readdir(FTPDIR);
closedir(FTPDIR);
chdir("c\:\\big\\seer2006");
foreach $datafile (@files)
  {
  open (TEXT, $datafile);
  $line = " ";
  while ($line ne "")
    {
    $line = <TEXT>;
    $code = substr($line, 47, 5);
```

```perl
    $code2 = substr($line, 52, 5);
    if (exists($dictionary{$code2}))
        {
        $code = $code2;
        }
    if (exists($dictionary{$code}))
        {
        if (substr($line,21,1) eq "0")   #0 means non-hispanic
            {
            next unless (substr($line,19,2) eq "02");
            $black_count++;
            next if ($code !~ /90[678]/);
            $bl_count{$code}++;
            }
        else
            {
            next if (substr($line,19,2) ne "01");
            $hisp_count++;
            next if ($code !~ /90[678]/);
            $hi_count{$code}++;
            }
        $germ_count++;
        }
    }
  close(TEXT);
  }
print "\nNumber of black persons with cancer is $black_count\n";
print "Number of hispanic persons with cancer is $hisp_count\n";
print "Total number of germ cell tumor records is $germ_count\n\n";
while((my $key, my $value) = each (%bl_count))
    {
    if ($bl_count{$key} > 5)
        {
        if ($hi_count{$key} > 5)
            {
            $hbfract = ($hi_count{$key} / $hisp_count) / ($bl_count{$key}
/ $black_count);
            printf ("%-3.02f %-35.35s\n", $hbfract, $dictionary{$key});
            }
        }
    }
exit;
```

Python Script

```python
#!/usr/local/bin/python
import sys, os, re, string
icd_in = open("c:/ftp/icdo3.txt", "r")
dictionary = {}
bl_count = {}
```

```
hi_count = {}
black_count = 0
hisp_count = 0
germ_count = 0
for line in icd_in:
  code_has = re.search(r'([0-9]{4})\/([0-9]{1}) +(.+)', line)
  if code_has:
    code = code_has.group(1) + code_has.group(2)
    term = code_has.group(3)
    term = string.rstrip(term)
    dictionary[code] = term
    bl_count[code] = 0
    hi_count[code] = 0
icd_in.close
filelist = os.listdir("c:/big/seer2006")
os.chdir("c:/big/seer2006")
for file in filelist:
  infile = open(file,'r')
  for line in infile:
    code = line[47:52]
    code2 = line[52:57]
    if dictionary.has_key(code2):
      code = code2
    hispanic_entry = re.match(r'[0-9]', line[21])
    if not hispanic_entry:
      continue
    race_entry = re.match(r'[012]', line[19])
    if not race_entry:
      continue
    if dictionary.has_key(code):
      if (line[21] == "0"):                        #0 means non-hispanic
        if not (line[19:21] == "02"):                #01 means white
          continue
        black_count = black count + 1
        germ_code_match = re.match(r'90[678]', code)
        if not germ_code_match:
          continue
        bl_count[code] = bl_count[code] + 1
      else:
        if not (line[19:21] == "01"):
          continue
        hisp_count = hisp_count + 1
        germ_code_match = re.match(r'90[678]', code)
        if not germ_code_match:
          continue
        hi_count[code] = hi_count[code] + 1
      germ_count = germ_count + 1
  infile.close()
os.chdir("c:/ftp/py")
```

```
print "Number of black persons with cancer is " + str(black_count)
print "Number of hispanic persons with cancer is " + str(hisp_count)
print "Total number of germ cell tumor records is " + str(germ_count)
print
for key,value in bl_count.iteritems():
  if (bl_count[key] > 5):
    if (hi_count[key] > 5):
      hbfract = (float(hi_count[key]) / float(hisp_count)) /
(float(bl_count[key]) / float(black_count))
      print "%-3.02f %-35.35s" % (hbfract, dictionary[key])
exit
```

Ruby Script

```ruby
#!/usr/local/bin/ruby
require 'mathn'
icd_in = File.open("c:/ftp/icdo3.txt", "r")
dictionary = {}
bl_count = Hash.new(0)
hi_count = Hash.new(0)
black_count = 0
hisp_count = 0
germ_count = 0
icd_in.each_line do
  |line|
  if line =~ /([0-9]{4})\/([0-9]{1}) +/
    code = $1 + $2
    term = $'
    term = term.sub(/ *\n/, "") if term =~ / *\n/
    dictionary[code] = term
  end
end
icd_in.close
filelist = Dir.glob("c:/big/seer2006/*.TXT")
filelist.each do
 |filepathname|
 seer_in = File.open(filepathname)
 seer_in.each_line do
   |line|
   code = line.slice(47, 5)
   code2 = line.slice(52, 5)
   if dictionary.has_key?(code2)
     code = code2
   end
   if dictionary.has_key?(code)
     if line.slice(21,1).eql?("0")  #0 means non-hispanic
       next unless line.slice(19,2).eql?("02")    #01 means white
       black_count = black_count + 1
       next if code !~ /90[678]/
       bl_count[code] = bl_count[code] + 1
```

```
      else
        next unless line.slice(19,2).eql?("01")
        hisp_count = hisp_count + 1
        next if code !~ /90[678]/
        hi_count[code] = hi_count[code] + 1
      end
      germ_count = germ_count + 1
    end
  end
  seer_in.close
end
puts "Number of black persons with cancer is " + black_count.to_s
puts "Number of hispanic persons with cancer is " + hisp_count.to_s
puts "Total number of germ cell tumor records is " + germ_count.to_s
puts
bl_count.each_pair do
  |key, value|
  if bl_count[key] > 5
     if hi_count[key] > 5
        hbfract = (hi_count[key].to_f / hisp_count.to_f) /
(bl_count[key].to_f / black_count.to_f)
        printf "%-3.02f %-35.35s\n", hbfract, dictionary[key]
     end
  end
end
exit
```

21.2 Analysis

Here is the output:

    Number of black persons with cancer is 318537
    Number of hispanic persons with cancer is 130155
    Total number of germ cell tumor records is 2100

The ratios of germ cell tumors in Hispanics compared with black persons are

    2.65 Germinoma
    7.08 Mixed germ cell tumor
    3.50 Spermatocytic seminoma
    7.10 Seminoma, anaplastic
    6.14 Seminoma, NOS
    3.44 Dysgerminoma
    1.81 Yolk sac tumor
    6.95 Teratocarcinoma
    8.61 Embryonal carcinoma, NOS
    1.99 Teratoma, malignant, NOS

There is a consistently higher incidence of cancer in the Hispanic population for every type of germ cell tumor. This tells us a few things. First, that all of the germ cell tumors are related to each other by more than histogenesis (cell of origin). They must have a relationship that extends to causation and development. Second, it tells us that the relatively high level of occurrence of germ cell tumors in white Hispanics is not just a fluke occurring in one cancer of one particular site. It is a consistent phenomenon that extends to all of the germ cell tumors.

Exercises

1. By adding three pound signs "#" to the script that compares germ cell occurrences in the Hispanic and black populations, you can produce a script that compares the occurrences of all tumors in these two populations. Where would you put the "#" signs?

2. The high rate of germ cell tumors in white Hispanics compared to that of black non-Hispanics may have one of two explanations: Hispanics may have a high rate of germ cell cancer; or, blacks may have a low rate of germ cell cancers. To resolve which possibility is correct, we could compare the relative frequencies of occurrence of germ cell tumors in several more subpopulations. If blacks have a consistently low frequency of occurrence of germ cell tumors, compared to several different populations, then we would be dealing with an unexplained low-risk population (rather than an unexplained high-risk population). Modify the script to produce relative occurrences of germ cell tumors in black persons compared with white non-Hispanic persons. Viewing this data, what can you conclude?

3. Modify the script to produce the same output, restricted to gender (male and female). You can do this with two separate scripts, or with one script that produces additional line items in the output file. What conclusions does this data provide? What additional hypotheses do these data provide?

4. Are the different rates of occurrence of germ cell tumors associated with differences in the age of occurrence of these tumors? Using Perl, Python, or Ruby, stratify the number of germ cell tumors occurring in white Hispanic and black groups, by age. You may divide ages into 10-year intervals, if you wish. What conclusions does this data provide? What additional hypotheses do these data provide?

22

CASE STUDY

Ranking the Death-Certifying Process, by State

As shown in Chapter 13, it is easy to parse a year's worth of deidentified death certificate data contained in one of the CDC public use mortality files.

We have been using the 1999 mortality file, which contains about 2.3 million records. Each record may list up to 20 diseases, representing the underlying and proximate causes of death and any significant additional conditions that the certifying doctor deems noteworthy. When physicians fill out a death certificate, they can choose to be thorough, by listing all diseases that lead to the patient's death, as well as other significant medical conditions affecting the patient at the time of death. Physicians can also choose to be somewhat lazy, listing the proximate cause of death, and omitting preceding and concurrent diseases. With a little effort, we can count the diseases listed in each cause of death record, and determine the average number of diseases reported for each state.

22.1 Script Algorithm

1. Because the CDC mortality files use a numeric code for each state, you must create a dictionary object whose keys are the code numbers for each state and whose values are corresponding names of states (or, in this case, the two-letter abbreviation for each state). The public file that explains the byte fields in the mortality files includes a state code listing, which we can extract as a simple text file, cdc_states.txt. This file is available at www.julesberman.info/book/cdc_states.txt. Parse through the cdc_states.txt file to build the state-by-state dictionary object.

2. Open the 1999 CDC mortality file.

3. Create a new dictionary object whose key will be the state abbreviation and whose value will be the total number of different records assigned to the state (i.e., the number of recorded deaths in the state).

4. Create a new dictionary object whose key will be the state abbreviation and whose value will be the total number of different cause of deaths diagnoses included in the aggregate of records for the state.

5. For each line record in the file, extract the section of the record that contains the various cause of death codes for the individual record (bytes 162 to 303).

6. For each line record in the file, extract the two-digit state code (bytes 21 and 22) and assign it the corresponding dictionary value (i.e., state abbreviation).
7. For each line record in the file, increment by 1 the dictionary object created in step 3 for the state specified in the record.
8. For each line record in the file, increment by 1 the dictionary object created in step 4 with the number of different conditions listed in the cause of death record. This is determined by the number of alphanumeric sequences that are separated by a space in bytes 162 to 303 of the record.
9. The rank of each state is determined by the total number of conditions recorded for all of the death certificate records for the state, divided by the total number of records for the state, and ranked by sorted numeric order for every state.

Perl Script

```perl
#!/usr/local/bin/perl
open (STATE, "cdc_states.txt");
$line = " ";
while ($line ne "")
   {
   $line = <STATE>;
   $line =~ /^[0-9]{2}/;
   $state_code = $&;
   $line =~ / +([A-Z]{2}) *$/;
   $state_abb = $1;
   $statehash{$state_code} = $state_abb;
   }
close STATE;
open (ICD, "Mort99us.dat");   #the CDC mortality file
$line = " ";
while ($line ne "")
   {
   $line = <ICD>;
   $codesection = substr($line,161,140);
   $code = substr($line,20,2);
   $state = $statehash{$code};
   $state_total{$state}++;
   $codesection =~ s/ +$//;
   $eager = scalar(split(" ",$codesection));
   $state_eager{$state} = $state_eager{$state} + $eager;
   }
while ((my $key, my $value) = each(%state_total))
   {
   $goodness = substr(($state_eager{$key} / $value),0,5);
   push(@list_array,  "$goodness $key");
   }
print join("\n", (sort(@list_array)));
exit;
```

Python Script

```python
#!/usr/local/bin/python
import re, string
state_file = open("c:\\ftp\\cdc_states.txt", "r")
state_hash = {}
state_eager = {}
state_total = {}
list_array = []
for line in state_file:
  codematch = re.match(r'([0-9]{2})', line)
  if codematch:
    state_code = codematch.group(1)
  statematch = re.search(r' +([A-Z]{2}) *$', line)
  if statematch:
    state_abb = statematch.group(1)
  state_hash[state_code] = state_abb
  state_total[state_abb] = 0
  state_eager[state_abb] = 0
state_file.close()
cdc_file = open("c:\\big\\Mort99us.dat", "r")
for line in cdc_file:
  codesection = line[161:300]
  code = line[20:22]
  state = state_hash[code]
  state_total[state] = state_total[state] + 1
  codesection = re.sub(r' +$', "", codesection)
  eager = re.split(r' +', codesection)
  eager = len(eager)
  state_eager[state] = state_eager[state] + eager
for key in state_total.keys():
  goodness = str(float(state_eager[key]) / state_total[key])
  goodness = goodness[0:5]
  list_array.append(goodness + " " + key)
list_array = sorted(list_array)
for item in list_array:
  print item
exit
```

Ruby Script

```ruby
#!/usr/local/bin/ruby
require 'mathn'
state_file = File.open("c:/ftp/cdc_states.txt")
state_hash = {}
state_eager = {}
state_total = {}
list_array = []
state_file.each_line do
```

```
  |line|
  line =~ /^[0-9]{2}/
  state_code = $&;
  line =~ / +([A-Z]{2}) *$/
  state_abb = $1;
  state_hash[state_code] = state_abb
end
state_file.close
cdc_file = File.open("c:/big/Mort99us.dat")
cdc_file.each_line do
  |line|
  codesection = line.slice(161,140)
  code = line.slice(20,2)
  state = state_hash[code]
  state_total[state] = state_total[state].to_i + 1
  codesection = codesection.sub(/ +$/, "")
  eager = codesection.split(" ").length
  state_eager[state] = state_eager[state].to_i + eager
end
state_total.each do
  |key, value|
  goodness = (state_eager[key] / value).to_f.to_s.slice(0,5)
  list_array.push(goodness + " " + key)
end
puts list_array.sort.join("\n")
exit
```

22.2 Analysis

How many diagnoses are typically listed on a death certificate? About three. Many certificates list only a single condition.

It is easy to rank the average number of conditions listed on the certificates, by state.

The lowest ranking state is AR (Arkansas), with an average of 2.442 conditions listed on each certificate. Next in line is Louisiana, with 2.47 conditions listed. Arizona follows with 2.501.

2.442 AR	2.631 IN
2.479 LA	2.632 WI
2.501 AZ	2.634 NM
2.531 AL	2.649 OR
2.554 MT	2.652 FL
2.567 MA	2.663 MI
2.579 NV	2.667 SD
2.603 OK	2.678 MN
2.603 VA	2.690 NJ
2.609 KY	2.714 UT
2.621 IL	2.768 AK

2.774 PA	2.903 DE
2.781 MS	2.909 NH
2.789 KS	2.921 NE
2.795 MO	2.935 DC
2.796 ID	2.949 NY
2.800 WY	2.955 MD
2.802 GA	2.956 CT
2.824 SC	3.083 ND
2.831 IA	3.102 WV
2.855 ME	3.125 VT
2.875 CO	3.138 OH
2.875 TX	3.195 RI
2.879 WA	3.316 HI
2.880 NC	3.363 CA
2.883 TN	

The highest-ranking state is California, with 3.363 conditions listed on each certificate. Next to the top is Hawaii, with 3.316 conditions.

What is a "lazy" death certificate? I would think that a lazy death certificate is one that contains the absolutely minimal number of conditions required to certify death (i.e., "1"). Let us rank the states by the fraction of death certificates, registered in the state, that contain only one listed condition for the cause of death (by tweaking the first script).

0.323 AL	0.244 AZ
0.304 MT	0.242 ID
0.303 AR	0.241 ME
0.291 KY	0.241 FL
0.290 IN	0.239 AK
0.288 LA	0.238 UT
0.285 MN	0.238 KS
0.277 VA	0.234 WA
0.274 WI	0.233 IA
0.270 SD	0.229 DE
0.267 MI	0.228 WY
0.267 IL	0.225 SC
0.258 PA	0.222 TN
0.255 OK	0.222 CO
0.255 MA	0.221 NC
0.252 OR	0.220 TX
0.249 NM	0.219 NV
0.249 NJ	0.217 DC
0.245 MO	0.214 MS

0.214 MD	0.180 RI
0.200 GA	0.177 VT
0.199 NH	0.176 CT
0.196 OH	0.171 HI
0.192 WV	0.129 NY
0.190 ND	0.119 CA
0.185 NE	•

Alabama has the worst performance, with nearly one-third of death certificates having only one listed condition. California, once more, has the best performance of all the states, with one condition reported in only about one-tenth of certificates (i.e., about 90% of certificates have more than one condition reported).

Just about every death involves multiple underlying causes of death leading to a proximate cause of death. The number of conditions listed on a death certificate is, in most cases, a matter of personal effort on the part of the certifying doctor.

It can be difficult to produce an accurate death certificate. Nonetheless, much of what we know about human disease and the causes of human mortality comes from examination of death certificates. Death certificates have profound importance for the family of the deceased. Doctors should be trained to provide complete and accurate entries for "causes of death" and "other significant conditions" on death certificates.

Exercises

1. Every death certificate must have at least one disease entry for the cause of death. Write a script that ranks each state by the percentage of death certificates that have only one disease listed in the cause of death section, as discussed in the chapter. The laziest state would be the state with the highest percentage of death certificates carrying only one disease listing for the cause of death.

2. Assuming that a death certificate is completed for every death occurring in the United States, write a script that determines the total number of deaths that occurred in the United States in 1999.

3. Determine whether there is a difference, based on ethnicity, between the number of diseases listed in the cause of death section of death certificates. Specifically, answer the question: "Do white persons, black persons, and Hispanic persons have about the same number of listed conditions on their respective death certificates?"

4. Death certificates, as we have seen, code diseases using ICD, a coding system that is entirely different from the ICD-O (the classification of cancers). Nonetheless, the ICD does include cancers which are all prefixed with a "C" in their code. For example, "C13.1" is the ICD code for a cancer occurring in the hypopharyngeal aspect of the aryepiglottic fold. Using Perl, Python, or Ruby, write a script that counts the number of death certificates that contain a cancer code (a code beginning with a "C") and determine the proportion of the total number of death certificates in the 1999 mortality file that include a cancer diagnosis.

23
CASE STUDY
Data Mashups for Epidemics

Data mashups combine and integrate different data sources to produce a graphical representation of data that could not be achieved with any single available data source. Many people apply the term *mashup* to Web-based applications that employ two or more Web services or that use two or more Web-based applications that have Web-accessible APIs (application program interfaces) that permit their data to be integrated into a derivative application. Because I am a biomedical information specialist, I apply the term *mashup* to any application that integrates available biomedical data sources to answer questions using a graphic output (with or without Web involvement).

The classic medical mashup was performed by Dr. John Snow in London in 1854. A major outbreak of cholera occurred in late August and early September of 1854, in the Soho district of London. By the end of the outbreak, 616 people died.

At the time, nobody understood the biological cause of cholera. At the height of the outbreak, Dr. Snow conducted a rapid, interview-based survey of the site of occurrences of new cases of cholera, producing a case-density map, hand-drawn by the doctor himself (Figure 23.1).

Examination of the map revealed that the epidemic expanded from a water source, the Broad Street pump. The pump was quickly shut. Dr. Snow's historic mashup is sometimes credited with ending the cholera epidemic and heralding a new age in scientific biomedical investigation.

Today, epidemiologic inferences are drawn from diverse data sources. As described by Donald McNeil for *The New York Times*, the track of an influenza epidemic, originating in Mexico might be predicted by following the flow of dollar bills between Mexico and the United States.[*]

To create a map mashup, we will need a data source that lists occurrences of disease and the localities in which they occur; a data source that provides the latitude and longitude of localities; and a map whose East, West, North, and South boundaries have known latitudes and longitudes.

---

[*] McNeil, D.G. Predicting Flu with the Aid of (George) Washington. The New York Times, May 3, 2009.

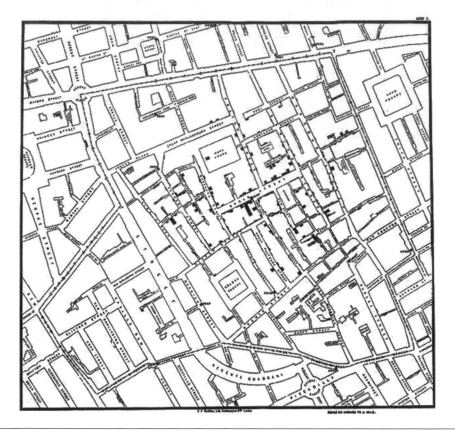

Figure 23.1   Cholera case occurrences in the 1854 London epidemic are shown in this early mashup of incidence data and geography; at the epicenter is the infamous Broad Street pump. (Author note: This map is now in the public domain, and a higher-resolution version of the map is available from Wikipedia at http://en.wikipedia.org/wiki/File:Snow-cholera-map-1.jpg.)

23.1 Tally of Coccidioidomycosis Cases by State

We can use the CDC mortality data set to create a mashup of disease occurrences in U.S. states.

Coccidioidomycosis (commonly misspelled Coccidiomycosis) is a fungus. Spores can lodge in the lungs of humans, producing debilitating and chronic pulmonary disease (Figure 23.2).

Let us examine the geographic distribution of coccidioidomycosis. We can write a short script that parses through every record in the CDC mortality file, pulling each death for which the diagnosis of coccidioidomycosis was recorded, and tallying the deaths for the states in which the deceased death certificate was recorded. This will tell us something about the state-by-state distribution of coccidioidomycosis.

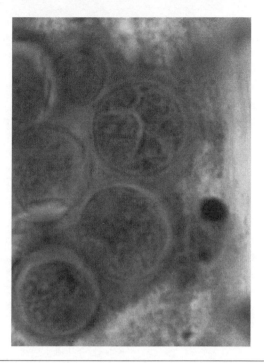

Figure 23.2 Histologic specimen of *Coccidioidomycosis immitis* in a sputum specimen.

23.1.1 Script Algorithm

1. Open the file, cdc_states.txt, which is the CDC's code list for each state and the District of Columbia (Figure 23.3). The two-digit state code corresponds to the "state" field in byte locations 21 and 22 of the U.S. mortality records.

2. Create a dictionary object whose keys are the state's two-digit codes and whose values are the corresponding state abbreviations. Use file available at www.julesberman.info/book/cdc_states.txt.

3. Open the 1999 U.S. mortality file (mort99us.dat), and parse through every line record.

4. As each line is parsed, extract the section of the record that contains the ICD codes for the conditions listed on the death certificates (bytes 160 to 300).

5. Match the regular expression containing the ICD code for coccidioidomycosis ("B38") against the record section containing the ICD codes.

6. If there is a match, determine the state in which the record occurred (byte 21 and 22) and increment a dictionary object whose keys are the state abbreviations and whose values are the cumulative tallies of the number of cases of coccidioidomycosis occurring in the state.

7. After the mortality file has been parsed, print out the key–value pairs from the dictionary object containing states and their coccidioidomycosis tallies.

01	Alabama AL	27	Montana MT
02	Alaska AK	28	Nebraska NE
03	Arizona AZ	29	Nevada NV
04	Arkansas AR	30	New Hampshire NH
05	California CA	31	New Jersey NJ
06	Colorado CO	32	New Mexico NM
07	Connecticut CT	33	New York NY
08	Delaware DE	34	North Carolina NC
09	District of Columbia DC	35	North Dakota ND
10	Florida FL	36	Ohio OH
11	Georgia GA	37	Oklahoma OK
12	Hawaii HI	38	Oregon OR
13	Idaho ID	39	Pennsylvania PA
14	Illinois IL	40	Rhode Island RI
15	Indiana IN	41	South Carolina SC
16	Iowa IA	42	South Dakota SD
17	Kansas KS	43	Tennessee TN
18	Kentucky KY	44	Texas TX
19	Louisiana LA	45	Utah UT
20	Maine ME	46	Vermont VT
21	Maryland MD	47	Virginia VA
22	Massachusetts MA	48	Washington WA
23	Michigan MI	49	West Virginia WV
24	Minnesota MN	50	Wisconsin WI
25	Mississippi MS	51	Wyoming WY
26	Missouri MO		

Figure 23.3   CDC's two-digit codes for the 50 states and the District of Columbia.

Perl Script

```
#!/usr/local/bin/perl
open (STATE, "cdc_states.txt");
$line = " ";
while ($line ne "")
  {
  $line = <STATE>;
  $line =~ /^[0-9]{2}/;
  $state_code = $&;
  $line =~ / +([A-Z]{2}) *$/;
  $state_abb = $1;
  $statehash{$state_code} = $state_abb;
  }
close STATE;
open (ICD, "Mort99us.dat");
$line = " ";
```

```perl
while ($line ne "")
   {
   $line = <ICD>;
   $state = 0;
   $codesection = substr($line,161,140);
   if ($codesection =~ /B38/)
      {
      $code = substr($line,20,2);
      $state = $statehash{$code};
      $state_tally{$state}++;
      }
   }
while ((my $key, my $value) = each(%state_tally))
   {
   print "$key $value\n";
   }
exit;
```

Python Script

```python
#!/usr/local/bin/python
import re, string
stat_in = open("c:/ftp/cdc_states.txt", "r")
statehash = {}
state_tally = {}
for line in stat_in:
   code_match = re.match(r'([0-9]{2})', line)
   if code_match:
      state_code = code_match.group(1)
   abb_match = re.search(r' +([A-Z]{2}) *$', line)
   if abb_match:
      state_abb = abb_match.group(1)
   statehash[state_code] = state_abb
   state_tally[state_abb] = 0
stat_in.close()
mort_in = open("c:/big/Mort99us.dat", "r")
for line in mort_in:
   state = 0
   codesection = line[161:302]
   cocc_match = re.search(r'B38', codesection)
   if cocc_match:
      code = line[20:22]
      state = statehash[code]
      state_tally[state] = state_tally[state] + 1
mort_in.close()
for key,value in state_tally.iteritems():
 if (value == 0):
   continue
 print key + " " + str(value)
exit
```

Ruby Script

```ruby
#!/usr/local/bin/ruby
stat_in = File.open("c:/ftp/cdc_states.txt")
statehash = {}
state_tally = Hash.new(0)
stat_in.each_line do
  |line|
  line =~ /^[0-9]{2}/
  state_code = $&
  line =~ / +([A-Z]{2}) *$/;
  state_abb = $1;
  statehash[state_code] = state_abb
end
stat_in.close
mort_in = File.open("c:/big/Mort99us.dat")
mort_in.each_line do
  |line|
  state = 0
  codesection = line.slice(161,140)
  if (codesection =~ /B38/)
    code = line.slice(20,2)
    state = statehash[code]
    state_tally[state] = state_tally[state] + 1
  end
end
mort_in.close
state_tally.each_pair {|key,value| puts key + " " + value.to_s}
exit
```

23.1.2 Analysis

The script produces a list of states and the tally of coccidioidomycosis cases, culled from the 1999 U.S. mortality file. Here is the output of the script:

Az 62	Mo 1	Pa 1
Ca 53	Mt 1	Tx 18
Id 2	Nc 2	Ut 2
Il 2	Nm 3	Wa 4
In 1	Nv 3	Wi 2
Ks 1	Ny 1	Wv 1
Ky 1	Oh 1	
Mn 1	Or 2	

You will notice that fewer than 50 states are included in the list. States that had no cases of coccidioidomycosis were not added to the list. We will see that this does not affect the mashup.

23.2 Creating the Map Mashup

Whenever we have a data set with numeric data associated with each state, we can use that data to create a mashup that projects the data onto a map of the United States. If the data has a recognizable geographic distribution, we will able to visualize the trend.

23.2.1 Script Algorithm

1. Import an image module into your script.
2. Open the external file, which contains the map coordinates for the geographic centers of each state. The file is available at the following location:

   http://www.maxmind.com/app/state_latlon

   State longitudes and latitudes, obtained from the state_latlon file, and used in this script, is available at

   http://www.julesberman.info/book/loc_states.txt

   For this script, we deposited loc_states.txt in the c:\ftp\ subdirectory of our hard drive.
   The first few lines of the file are shown here, with the latitude and longitudes for Alaska, Alabama, and Arkansas.

   "AK,61.3850,-152.2683"
   "AL,32.7990,-86.8073"
   "AR,34.9513,-92.3809"

   These three lines mean the following:

   Alaska Latitude 61.3850 (North) Longitude 152.2683 (West)
   Alabama Latitude 32.7990 (North) Longitude 86.8073 (West)
   Arkansas Latitude 34.9513 (North) Longitude 92.3809 (West)

3. Create three dictionary objects. In both objects, the two-letter state codes are the keys. In one, the values are the latitude locations of the states. In the other, the values are the longitude locations of the states. In the third, the values are the number of cases of coccidioidomycosis occurring in the state (data prepared in the prior section).
4. Open an image file consisting of a map of the United States. We will use the same U.S. map that we used in Chapter 3 to "mash up" the disease data.

   http://www.julesberman.info/book/us.gif

5. The coordinates of the perimeter of the map are as follows:

   north = 49°; #Northernmost latitude of map in degrees north
   south = 25°; #Southernmost latitude of map in degrees north

west = 125°; #Westernmost longitude of map in degrees west

east = 66°; #Easternmost longitude of map in degrees west

The location of each state can be positioned to a specific point on the map by calculating the fraction of the map's north-south and east-west distances (in degrees) that is occupied by each state's latitude and longitude.

6. Determine the number of columns and rows in the map image. This gives you the width (columns) and height (rows) of the full image.

7. For each state, translate the global coordinates for each state as *x,y* coordinates on the map image.

8. Draw circles on the map, using the *x,y* coordinates for each state as the center for each circle, and using the number of cases of coccidioidomycosis (for each state) for the proportionate size of the radius of the circle for the state.

9. After circles are drawn for each state, write the resulting image to an external image file.

Perl Script

```perl
#!/usr/local/bin/perl
use Image::Magick;
$north = 49;
$south = 25;
$west = 125;
$east = 66;
open(TEXT, "c\:\\ftp\\loc_states.txt");
$line = " ";
while ($line ne "")
    {
    $line = <TEXT>;
    $line =~ /^([A-Z]{2})\,([0-9\.]+)\,\-?([\.0-9]+) *$/;
    $state = $1;
    $latitude = $2;
    $longitude = $3;
    $lathash{$state} = $latitude;
    $lonhash{$state} = $longitude;
    }
close(TEXT);
open(TEXT, "c\:\\ftp\\state_count.txt");
$line = " ";
while ($line ne "")
    {
    $line = <TEXT>;
    $line =~ / /;
    $state_abb = $`;
    $state_value = $';
    $sizehash{$state_abb} = $state_value;
    }
my $img1 = Image::Magick->new;
```

```perl
$img1 -> ReadImage("c\:\\ftp\\us.gif");
$width = $img1 -> Get('columns');
$height = $img1 -> Get('rows');
while ((my $key, my $value) = each(%lathash))
    {
    $state = $key;
    $latitude = $value;
    $longitude = $lonhash{$key};
    $offset_y =  int((($north - $latitude) / ($north - $south)) *
$height);
    $offset_x =  int((($west - $longitude) / ($west - $east)) *
$width);
    if (exists($sizehash{$key}))
      {
      $radius = $offset_x + (2 * $sizehash{$key});
      }
    else     #for the states with no cases of coccidioidomycosis
      {
      $radius = $offset_x + 0;
      }
    $img1 -> Draw (
      stroke    => "red",
      primitive => "circle",
      points    => "${offset_x}, ${offset_y}, ${radius},
${offset_y}");
    }
$img1 -> write ("gif:cocc_pl.gif");
exit;
```

Python Script

```python
#!/usr/local/bin/python
import sys
import Image, ImageDraw
import re
lathash = {}
lonhash = {}
sizehash = {}
north = 49
south = 25
west = 125
east = 66
infile = open ("loc_states.txt", "r")
for line in infile:
    match_tuple = re.match(r'^([A-Z]{2})\,([0-9\.]+)\,\-?([\.0-9]+)
*$',line)
    state = match_tuple.group(1)
    latitude = float(match_tuple.group(2))
    longitude = float(match_tuple.group(3))
    lathash[state] = latitude
```

```
      lonhash[state] = longitude
      sizehash[state] = 0
infile.close()
infile = open("c:/ftp/state_count.txt", "r")
for line in infile:
      space_match = re.match(r'([A-Z]+) ([0-9]+)$', line)
      if space_match:
         state_abb = space_match.group(1)
         state_value = space_match.group(2)
         sizehash[state_abb] = state_value
infile.close()
im = Image.open("c:/ftp/us.jpg", "r")
print im.mode
[width, height] = im.size
draw = ImageDraw.Draw(im)
for state, latitude in lathash.iteritems():
   longitude = lonhash[state]
   offset_y =  int(((north - latitude) / (north - south)) * height)
   offset_x =  int(((west - longitude) / (west - east)) * width)
   radius = 2 * int(sizehash[state])
   draw.ellipse((offset_x, offset_y, (offset_x + radius), (offset_y
+ radius)), outline=0xff0000, fill=0x0000ff)
im.save("cocc_py.jpg")
exit
```

Ruby Script

```
#!/usr/local/bin/ruby
require 'RMagick'
north = 49.to_f #degrees latitude
south = 25.to_f #degrees latitude
west = 125.to_f #degrees longitude
east = 66.to_f  #degrees longitude
#corresponds to the continental perimeters
text = File.open("c\:\\ftp\\loc_states.txt", "r")
lathash = Hash.new
lonhash = Hash.new
text.each do
    |line|
    line =~ /^([A-Z]{2})\,([0-9\.]+)\,\-?([\.0-9]+) *$/
    state = $1
    latitude = $2
    longitude = $3
    lathash[state] = latitude.to_f
    lonhash[state] = longitude.to_f
end
text.close
text = File.open("c\:\\ftp\\state_count.txt", "r")
sizehash = Hash.new
text.each do
```

```
  |line|
  line =~ / /
  state_abb = $`
  state_value = $'
  sizehash[state_abb] = state_value
end
text.close
img1 = Magick::ImageList.new("c\:\\ftp\\us\.gif")
width = img1.columns
height = img1.rows
gc = Magick::Draw.new
lathash.each do
  |key,value|
  state = key
  latitude = value.to_f
  longitude = lonhash[key].to_f
  l_y =  (((north - latitude) / (north - south)) * height).ceil
  l_x =  (((west - longitude) / (west - east)) * width).ceil
  gc.fill_opacity(0)
  gc.stroke('red').stroke_width(1)
  circlesize = ((sizehash[state].to_f)*2).to_i
  gc.circle(l_x, l_y, (l_x - circlesize), l_y)
  gc.fill('black')
  gc.stroke('transparent')
  gc.text((l_x - 5), (l_y + 5), state)
  gc.draw(img1)
end
img1.border!(1,1, 'lightcyan2')
img1.write("cocc_rb.gif")
exit
```

23.2.2 Analysis

The result is shown in Figure 23.4.

It took under a minute to generate this mashup map, parsing over one gigabyte of deidentified death certificate data records, extracting data on occurrences of both diseases and the states in which they were recorded, and producing a visual output that conveys a detailed epidemiologic study that can be understood at a glance.

We showed how we can use the CDC mortality data set to create a mashup, using short scripts. We started with a blank outline map of the United States, and we ended with a map indicating the occurrences of coccidioidomycosis in each state. Each state has been "pasted" into the U.S. map. States with red circles contained cases of coccidioidomycosis recorded on death certificates; the diameter of circles is proportionate to the number of cases.

At a glance, we can see that coccidioidomycosis occurs primarily in the Southwest United States. In fact, coccidioidomycosis, variously known as valley fever, San Joaquin

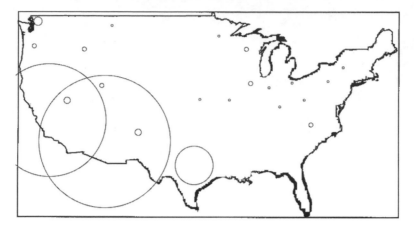

Figure 23.4   Recorded death certificate occurrences of coccidioidomycosis, by state.

Valley fever, California valley fever, and desert fever, is a fungal disease caused by *Coccidioides immitis*. In the United States, this disease is endemic to certain parts of the Southwest.

Exercises

1. In the script that calculated the occurrences of coccidioidomycosis in each state, only states with documented cases were included in the output. Add one line (or less) to the script (in Perl, Python, or Ruby) to produce an output that lists all of the states, assigning zero as the occurrence number for the states that had no coccidioidomycosis conditions on death certificates.

2. The output of our mashup script is an image file that we can view in any image viewing application. Revise the mashup script to automatically display the image file, once it is created. (*Hint:* Append the image-viewing code developed in Chapter 3.)

3. In the script that built the mashup map from the state-by-state data for coccidioidomycosis, we could have easily added data for other diseases, assigning different colors to the disease "circles" marking each disease in each state. Histoplasmosis is endemic in the eastern half of the United States. Conveniently, ICD codes for conditions resulting from *C. immitis* infection (coccidioidomycosis) all begin with B38. Conditions for *H. capsulatum* (histoplasmosis) begin with B39. Using Perl, Python, or Ruby, tweak the two prior scripts (to compute the occurrences of histoplasmosis in U.S. states, and to superimpose the occurrences of histoplasmosis onto the map of coccidioidomycosis occurrences. The output file should easily distinguish the geographic distribution of coccidioidomycosis and histoplasmosis (Figure 23.5).

4. So far, we have represented the number of occurrences of coccidioidomycosis in the different states. State-by-state raw occurrences of diseases can be misleading. States with large populations (such as California) may seem to have a much higher incidence of disease than a state with fewer people (such as Montana). To avoid this problem, occurrences are usually expressed as a rate (the number of cases in a state divided by the population of the state). For this,

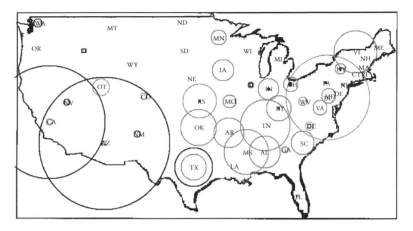

Figure 23.5 The geographic distributions of coccidioidomycosis and histoplasmosis. The black circles represent the coccidioidomycosis rates of infection, as recorded on death certificates. The gray circles represent histoplasmosis rates.

we can use a table that provides the population of each state in 1999, and we can represent the occurrence count as the total number of occurrences divided by the population of the state, multiplied by 100,000 (giving us the occurrences of the disease per 100,000 population).

Our data source was the U.S. mortality files for 1999. The CDC provides supplemental data for the mortality files by anonymous ftp, from ftp.cdc.gov, in the following subdirectory:

/pub/Health_Statistics/NCIIS/Dataset_Documentation/mortality/
    Mort99doc.pdf

This document provides the state populations (Figure 23.6).

The state population data from the Mort99doc.pdf file is available in a simple text file that you can download and use in a script:

http://www.julesberman.info/book/statepop.txt

Using Perl, Python, or Ruby, revise the script that calculates the rate of occurrences of coccidioidomycosis in each state (number of occurrences divided by the state population multiplied by 100,000) and displays the distribution in a mashup map.

Table L. Estimated population, by age, for the United States, each division and State, Puerto, Rico,
Virgin Islands, Guam, American Samoa, and Northern Marianas: July 1, 1999
[Figures include Armed Forces stationed in each area, and exclude Armed Forces stationed outside the United States]

Area	Total	Under 1 year	1–4 years	5–14 years	15–24 years	25–34 years	35–44 years	45–54 years	55–64 years	65–74 years	75–84 years	85 years and over
United States.....	272,690,813	3,819,903	15,122,239	39,495,230	37,773,512	37,935,812	44,812,649	35,802,358	23,389,085	18,218,248	12,146,695	4,175,082
Alabama	4,369,862	59,792	231,041	590,148	624,730	614,609	692,778	582,175	406,637	310,568	192,412	64,972
Alaska	619,500	9,703	40,062	113,329	104,654	72,283	106,349	91,254	47,116	21,975	10,377	2,398
Arizona	4,778,332	76,688	309,300	739,189	669,065	628,940	735,990	588,143	402,384	341,024	221,704	65,905
Arkansas	2,551,373	35,799	141,850	366,129	367,919	327,575	378,711	326,205	245,843	191,328	125,515	44,499
California	33,145,121	503,227	1,996,031	5,058,628	4,684,221	5,114,990	5,592,337	4,107,384	2,440,771	1,930,889	1,292,566	424,077
Colorado	4,056,133	58,422	229,793	592,039	577,959	523,975	698,674	607,548	359,950	221,673	138,479	47,621
Connecticut	3,282,031	42,707	175,458	478,846	386,963	446,479	563,663	436,112	283,227	231,909	173,456	63,211
Delaware	753,538	10,299	39,866	102,238	99,302	113,221	130,219	97,062	63,196	54,477	33,517	10,141
District of Columbia	519,000	5,932	21,368	54,840	58,821	95,007	88,718	72,725	49,487	38,181	24,698	9,223
Florida	15,111,244	190,737	761,637	2,033,258	1,820,203	1,881,169	2,357,168	1,891,468	1,433,755	1,429,984	991,262	320,603
Georgia	7,788,240	119,066	461,084	1,140,252	1,110,401	1,205,249	1,337,846	1,023,436	629,763	419,257	256,548	85,338
Hawaii	1,185,497	16,742	63,645	162,143	166,543	146,817	198,212	165,659	103,847	88,018	56,576	17,295
Idaho	1,251,700	18,863	73,972	192,032	209,572	154,208	186,707	163,298	111,019	73,510	50,592	17,927
Illinois	12,128,370	176,578	701,101	1,783,938	1,662,918	1,701,968	2,002,805	1,569,666	1,033,219	771,168	532,621	192,388
Indiana	5,942,901	82,957	330,718	850,473	841,153	823,952	962,925	784,770	522,933	391,285	261,480	90,255
Iowa	2,869,413	36,380	146,440	402,039	417,004	356,641	440,536	379,687	262,199	207,766	156,221	64,500
Kansas	2,654,052	37,382	146,631	387,725	398,281	340,484	427,182	342,695	219,593	175,171	127,266	51,642
Kentucky	3,960,825	52,621	206,472	534,114	576,942	542,574	640,592	541,556	372,800	267,645	168,067	57,442
Louisiana..........	4,372,035	64,438	249,712	659,054	697,815	571,970	674,905	568,266	384,417	277,259	168,291	55,908
Maine	1,253,040	13,439	53,779	167,703	166,148	166,472	218,280	180,810	111,052	92,615	60,561	22,181
Maryland	5,171,634	69,852	277,006	749,685	654,867	760,262	937,609	699,077	426,315	321,639	208,861	66,461

Figure 23.6 U.S. states, with total populations, and populations stratified by age. This is a portion of Table L from the publicly available file, Mort99doc.pdf.

24

CASE STUDY

Sickle Cell Rates

In 1949, Linus Pauling and co-workers showed that sickle cell anemia is a disease produced by an inherited alteration in hemoglobin, producing a molecule that is separable from normal hemoglobin by electrophoresis. Electrophoresis is still used to distinguish sickle hemoglobin from normal hemoglobin.

In 1956, Vernon Ingram and J.A. Hunt sequenced the hemoglobin protein molecule (normal and sickle cell) and showed that the inherited alteration in sickle cell hemoglobin is due to a single amino acid substitution in the protein sequence.

Because sickle cell hemoglobin can be detected by a simple blood test, it was assumed, back in the 1950s, that new cases of this disease would be prevented through testing, followed by genetic counseling. Today, there are a number of private and public organizations that work to reduce the incidence of sickle cell disease.

I have been interested in knowing whether sickle cell incidence is decreasing in the U.S. population. Despite PubMed and Web searches, I have not been able to find a single data source on the subject. Let us investigate using the CDC (U.S. Centers for Disease Control and Prevention) mortality data sets. We will need the CDC mortality files for the years 1996, 1999, 2002, and 2004, all of which contain deidentified records listing multiple conditions, coded in ICD10 (International Classification of Disease, version 10), for the underlying causes of death and other significant conditions, found on U.S. death certificates.

24.1 Script Algorithm

1. Information for obtaining the free, public use, CDC mortality files is available in the appendix. Put the filenames of the CDC mortality files into an array:

   mort96us.dat, mort99us.dat, mort02us.dat, mort04us.dat

2. Parse through each mortality file, line by line (about 5 GB total).
3. Records in the mortality files comprise a single line, with the list of death certificate conditions concatenated in 140 bytes, starting from line byte 449 (in the 1996 mortality file), 162 (in the 1999 mortality file), 163 (2002 mortality file), or 165 (2004 mortality file).
4. Extract the 140-byte section of each record, and match the conditions listed against the codes for sickle-cell-related diseases.

5. In the ICD, sickle cell disease and all variants of the disease begin with "D57". No other diseases have a code that begins with "D57".

D57 Sickle cell disorders
D57.0 Sickle cell anemia with crisis
D57.00 with crisis (vasoocclusive pain)
D57.01 with - - - - acute chest syndrome
D57.02 splenic sequestration
D57.1 Sickle cell anemia without crisis
D57.2 Double heterozygous sickling disorders
D57.20 without crisis
D57.211 with - - - - - acute chest syndrome
D57.212 splenic sequestration
D57.219 with crisis (vasoocclusive pain)
D57.3 Sickle cell trait
D57.40 without crisis
D57.411 with - - - - - acute chest syndrome
D57.412 splenic sequestration
D57.419 with crisis (vasoocclusive pain)
D57.8 Other sickle cell disorders
D57.80 spherocytosis
D57.811 with - - - - - acute chest syndrome
D57.812 splenic sequestration
D57.819 with crisis

If a death certificate indicates that a person died with sickle cell disease, the string sequence "D57" will occur somewhere in the 140-byte sequence extracted from the record.

6. Keep a tally of the occurrences of sickle cell conditions and the rate of occurrences (occurrences as a fraction of the population) for the available years (1996, 1999, 2002, and 2004).

Perl Script

```
#!/usr/local/bin/perl
@filearray = qw(mort96us.dat mort99us.dat mort02us.dat mort04us.dat);
foreach $file (@filearray)
{
open (ICD, $file);
$line = " ";
$popcount = 0;
$counter = 0;
while ($line ne "")
  {
  $line = <ICD>;
```

```
   $codesection = substr($line,448,140) if ($file eq "mort96us.dat");
   $codesection = substr($line,161,140) if ($file eq "mort99us.dat");
   $codesection = substr($line,162,140) if ($file eq "mort02us.dat");
   $codesection = substr($line,164,140) if ($file eq "mort04us.dat");
   $popcount++;
   if ($codesection =~ /D57/i)
     {
     $counter++;
     }
   }
close ICD;
$rate = $counter / $popcount;
$rate = substr((100000 * $rate),0,5);
print "\n\nDeath certificates listing sickle cell\n";
print " disease is $counter in $file file";
print "\nDeath certificate rate of sickle cell disease\n";
print "is $rate per 100,000 in $file file";
}
exit;
```

Python Script

```python
#!/usr/local/bin/python
import re
sickle_match = re.compile('D57')
lst = ("mort96us.dat","mort99us.dat","mort02us.dat","mort04us.dat")
for file in lst:
  intext = open(file, "r")
  popcount = 0
  counter = 0
  codesection = ""
  for line in intext:
    if file == lst[0]:
      codesection = line[448:588]
    if file == lst[1]:
      codesection = line[161:301]
    if file == lst[2]:
      codesection = line[162:302]
    if file == lst[3]:
      codesection = line[164:304]
    popcount = popcount + 1
    p = sickle_match.search(codesection)
    if p:
      counter  = counter + 1
  intext.close
  rate = float(counter) / float(popcount) * 100000
  rate = str(rate)
  rate = rate[0:5]
  print ('\n\nRecords listing sickle cell is ')
  print (str(counter) + ' in ' + file + ' file')
```

```
  print ('\nSickle cell rate per 100,000 records is ')
  print(str(rate) + ' in ' + file + ' file')
exit
```

Ruby Script

```ruby
#!/usr/local/bin/ruby
filearray = Array.new
filearray = "mort96us.dat mort99us.dat mort02us.dat mort04us.dat".
split
filearray.each do
  |file|
  text = File.open(file, "r")
  counter = 0; popcount = 0;
  text.each_line do
    |line|
    codesection = line[448,140] if (file == filearray.fetch(0))
    codesection = line[161,140] if (file == filearray.fetch(1))
    codesection = line[162,140] if (file == filearray.fetch(2))
    codesection = line[164,140] if (file == filearray.fetch(3))
    popcount = popcount +1
    counter = (counter + 1) if (codesection =~ /D57/i)
    end
  text.close
  rate = ((counter.to_f / popcount.to_f) * 100000).to_s[0,5]
  puts "\nRecords listing sickle cell is #{counter} in #{file} file"
  puts "Sickle cell rate per 100,000 records is #{rate} in #{file}
file"
  end
exit
```

24.2 Analysis

The script that parses through about 5 GB of CDC records and compiles the following results:

In 1996, U.S. cases with sickle cell disease in death certificates is 708
In 1996, U.S. rate of sickle cell disease in death certificates is 30.54 per 100,000
In 1999, U.S. cases with sickle cell disease in death certificates is 799
In 1999, U.S. rate of sickle cell disease in death certificates is 33.36 per 100,000
In 2002, U.S. cases with sickle cell disease in death certificates is 827
In 2002, U.S. rate of sickle cell disease in death certificates is 33.79 per 100,000
In 2004, U.S. cases with sickle cell disease in death certificates is 876
In 2004, U.S. rate of sickle cell disease in death certificates is 36.47 per 100,000

For all four years examined, there has been a steady, increasing trend in the number of death certificates listing sickle cell disease as a cause of death or a significant condition at the time of death. Likewise, the overall rate (per 100,000 certificates) has steadily increased in every sampled year, covering 1996 to 2004.

Does this mean that efforts to reduce the incidence of sickle cell disease have failed? Not necessarily. Death certificate data is unreliable. Whether a doctor thinks of adding sickle cell disease as a medical condition on the death certificate may depend on a variety of factors (as discussed previously). However, when you are dealing with very large numbers, trends usually reflect reality.

The best data would be natality incidence rates, by year, measured between about 1960 and the present. However, I have not been able to find that kind of data, and the CDC mortality files may be the next-best option.

Exercises

1. Using Perl, Python, or Ruby, modify the script to examine the trends in the occurrence of thalassemia in the CDC mortality files. The ICD codes for thalassemia are
   D56 Thalassemia
   D56.0 Alpha thalassemia
   D56.1 Beta thalassemia
   D56.2 Delta-beta thalassemia
   D56.3 Thalassemia trait
   D56.4 Hereditary persistence of fetal hemoglobin [HPFH]
   D56.8 Other thalassemias
   D56.9 Thalassemia, unspecified

2. Using Perl, Python, or Ruby, determine the trends for the occurrence of sickle cell anemia by state. Are there some states where the occurrences of sickle cell anemia (on death certificates) are dropping? (*Hint:* Use cdc_states.txt.)

3. Using Perl, Python, or Ruby, determine the average age at death of patients whose death certificates include a diagnosis of sickle cell anemia.

4. Examine the file that lists all of the diseases coded by the ICD (each10.txt).
   This file is available by anonymous ftp from the ftp.cdc.gov Web server in the following subdirectory:

   /pub/Health_Statistics/NCHS/Publications/ICD10/each10.txt

   or at

   http://www.julesberman.info/book/each10.txt

   Choose a condition in which you have a particular interest, and repeat the analysis from this chapter, substituting your chosen disease for sickle cell anemia.

25

CASE STUDY

Site-Specific Tumor Biology

We commonly speak of tumors as a general concept that has no specific anatomic site. Squamous cell carcinomas are tumors of squamous cells, and they arise wherever squamous cells happen to grow. The assumption is that squamous cell carcinomas all have the same general biological properties wherever they arise. The same is true for just about any tumor. A mesothelioma is a tumor that arises from the lining cells of body cavities. We might expect mesotheliomas to have the same biological properties wherever they might arise.

The problem with this way of thinking is that it depends on a large number of assumptions about the early development, growth, and treatment of tumors of the same name arising in different locations. Do we really know that mesotheliomas of the pleura are caused by the same agents that cause mesotheliomas of the peritoneum? Do we know that mesotheliomas of pleura occur in the same population of people as mesothelioma of the peritoneum? Might the cells of origin of pleural mesotheliomas have a different set of properties compared with the cells of origin of peritoneal mesotheliomas?

In prior chapters, we learned that different diseases occur with different frequencies in different ethnic or racial subpopulations. In this chapter, we will examine a single instance of a neoplasm that can occur in several different sites in the human body, and we will ask whether the tumors arising in different sites will occur in different numbers, and in different age distributions.

25.1 Anatomic Origins of Mesotheliomas

During embryonic development, cavities arise in the mesoderm (the middle embryonic layer), which eventually become the major body cavities, the pleura and the peritoneum. These cavities are lined by flat cells of mesodermal origin that produce small quantities of lubricating fluid rich in hyaluronic acid and chondroitin sulfate. The viscera of the chest (heart and lungs) hang within the pleural cavity (Figure 25.1).

The viscera of the abdomen (intestines, liver) are suspended wholly or partly within the peritoneal cavity (Figure 25.2). Some organs lie beneath the pleura or the peritoneum, and others lie in cavity recesses (the ovaries and para-ovarian tissues, and the testes and para-testicular tissues).

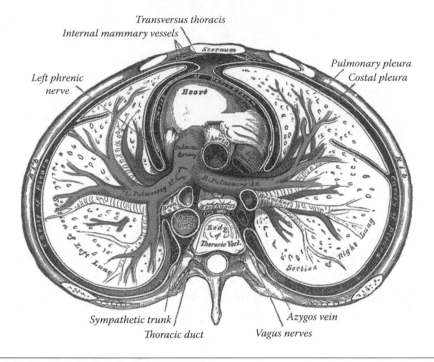

Figure 25.1   Pleural cavity (from *Gray's Anatomy*, 1918).

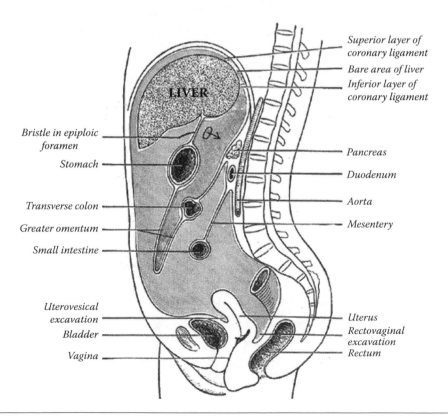

Figure 25.2   Peritoneal cavity (from *Gray's Anatomy*, 1918).

Mesotheliomas are malignant tumors that arise from the surfaces lining the walls of the body cavities, and from the surfaces of organs that lie within the body cavities.

25.2 Mesothelioma Records in the SEER Data Sets

In the SEER data sets, records of mesotheliomas are coded with one of four different ICD-O codes:

90503 = mesothelioma
90513 = fibrous mesothelioma
90523 = epithelial mesothelioma
90533 = biphasic mesothelioma

SEER records also contain a so-called topography code indicating the anatomic site from which the tumor arose. When we evaluate the topographic locations from which pleural and peritoneal mesotheliomas arise, we need to include topographic codes for the pleural and peritoneal cavities, as well as the organs within those cavities. For this exercise, we will lump together tumors that arise from the walls of the cavities (the parietal mesothelium) and tumors that arise from the surfaces of viscera within body cavities (the visceral mesothelium). A subset of mesotheliomas can arise from genital organs (ovaries and testes) found in recesses of the peritoneum, and we will collect separately those SEER records that account for mesotheliomas of ovary or testis.

The topography values are

C341 Upper lobe, lung
C342 Middle lobe, lung
C343 Lower lobe, lung
C348 Overlapping lesion of lung
C349 Lung NOS
C380 Heart
C381 Anterior mediastinum
C382 Posterior mediastinum
C383 Mediastinum NOS
C384 Pleura NOS
C388 Overlapping lesion of heart, mediastinum, and pleura
C390 Upper respiratory tract, NOS
C398 Overlapping lesion of respiratory system and intrathoracic organs
C399 Respiratory tract, NOS
C482 Peritoneum NOS
C488 Overlapping lesion of retroperitoneum and peritoneum
C540 Isthmus uteri
C541 Endometrium
C542 Myometrium

C543 Fundus uteri

C548 Overlapping lesion of corpus uteri

C549 Corpus uteri

C559 Uterus NOS

C569 Ovary

C570 Fallopian tube

C571 Broad ligament

C572 Round ligament

C573 Parametrium

C574 Uterine adnexa

C577 Wolffian body

C578 Overlapping lesion of female genital organs

C579 Female genital tract, NOS

C620 Undescended testis (site of neoplasm)

C621 Descended testis

C629 Testis NOS

C630 Epididymis

C631 Spermatic cord

C632 Scrotum, NOS

C637 Tunica vaginalis

C638 Overlapping lesion of male genital organs

C639 Male genital organs, NOS

25.2.1 Script Algorithm

1. Parse through SEER files, line by line. Each line of a SEER file is the record of a cancer occurrence, and there are over 3.7 million lines that will be parsed. Instructions for obtaining the free, public use SEER files are found in the appendix. In this script, the SEER files happen to reside in my computer's c:\ftp\seer2006 subdirectory.

2. As each line of the file is parsed, extract the 5-character substring that begins at character 48 and the 5-character substring that begins at character 53. These represent the ICD-O code for the record. The string beginning at character 53 is the code for the newer version of ICD-O.

3. As we saw in a prior section of this chapter, the code numbers for mesothelioma cases all begin with the sequence "905". If neither of the old record's ICD-O codes begin with the sequence "905", the record does not contain a mesothelioma, and the parsing loop can begin a new iteration, beginning with the next line.

4. For the SEER records that contain a mesothelioma diagnosis, keep a running tally of the total number of mesotheliomas.

5. For the SEER records that contain a mesothelioma diagnosis, determine the topographic code for the record. The topographic code corresponds to a four-character string sequence beginning at character 43 of the record. The topographic codes that interest us can be described with the following five regex expressions:

C34[12389]—pleural tumors
C3[89][0123489]—pleural tumors
C48[28]—peritoneal tumors
C5[4567]—ovarian tumors
C6[23]—testicular tumors

6. Capture the location of the mesothelioma as a variable, that will be one of the following types:

mesothelioma_pleura
mesothelioma_peritoneum
mesothelioma_ovary
mesothelioma_testis

7. As each line of the file is parsed, extract the three-digit number representing the age of the patient at the time that the tumor was diagnosed. The age is found in record characters 25, 26, and 27 (i.e., the 24th, 25th, and 26th characters when character 1 is counted as the 0th character).

8. Bin each age into one of 20 bins by dividing the age by 5, taking the integer value of the result, and lumping all ages 95 and above to the same bin (the 20th bin).

9. For each record, increment (by 1) the number of occurrences of the diagnosis (for the record), in the bin corresponding to the patient age listed for the record, and save the record diagnosis and the incremented age distribution for the diagnosis, as a key–value pair in a dictionary object.

10. Build a dictionary object consisting of the four anatomic sites of mesotheliomas as keys and the incremented age distributions as the values.

11. After the SEER files have been parsed, print out the key–value pairs of the dictionary object containing the anatomic sites of the mesotheliomas as keys and the age distributions of the occurrences of mesotheliomas as the values.

12. Print out the variable containing the tally of the total number of mesothelioma records in the SEER data sets.

Perl Script

```
#!/usr/local/bin/perl
opendir(SEERDIR, "c\:\\big\\seer2006") || die ("Unable to open
directory");
@files = readdir(SEERDIR);
closedir(SEERDIR);
```

```perl
chdir("c\:\\big\\seer2006");
foreach $datafile (@files)
  {
  next if ($datafile !~ /.txt/i);
  open (TEXT, $datafile);
  $line = " ";
  while ($line ne "")
    {
    $line = <TEXT>;
    $dx = substr($line, 47, 5);
    $dx2 = substr($line, 52, 5);
    next unless ($dx =~ /^905/ || $dx2 =~ /^905/);
    $count++;
    $place = substr($line,42,4);
    if ($place =~ /C34[12389]/)
      {
      $place = "pleura";
      }
    elsif ($place =~ /C3[89][0123489]/)
      {
      $place = "pleura";
      }
    elsif ($place =~ /C48[28]/)
      {
      $place = "peritoneum";
      }
    elsif ($place =~ /C5[4567]/)
      {
      $place = "ovary";
      }
    elsif ($place =~ /C6[23]/)
      {
      $place = "testis";
      }
    else
      {
      next;
      }
    $dxp = "mesothelioma_" . $place;
    unless (exists($dxhash{$dxp}))
      {
      $dxhash{$dxp} = "0 0 0 0 0 0 0 0 0 0 0 0 0 0 0 0 0 0 0 0 0 0";
      }
    $age_at_dx = substr($line,24,3);
    if ($age_at_dx > 95)
      {
      $age_at_dx = 95;
      }
    $age_at_dx = int($age_at_dx / 5);
```

```
    @agearray = split(" ", $dxhash{$dxp});
    $agearray[$age_at_dx]++;
    $dxhash{$dxp} = join(" ", @agearray);
    }
  close TEXT;
  }
while (($key, $value) = each(%dxhash))
  {
  @value_array = split(/ /, $value);
  print "$key\|$value\n";
  }
print "Total number of mesotheliomas is $count\n";
exit;
```

Python Script

```python
#!/usr/local/bin/python
import os, re, string
count = 0
dxhash = {}
filelist = os.listdir("c:/big/seer2006")
os.chdir("c:/big/seer2006")
for file in filelist:
  infile = open(file,'r')
  for line in infile:
    code = line[47:52]
    code2 = line[52:57]
    meso_match = re.match(r'905', code)
    if not meso_match:
      meso_match = re.match(r'905', code2)
      if not meso_match:
        continue
    count = count + 1
    place = line[42:46]
    pleura_yes = re.match(r'C34[12389]', place)
    pleura_yes = re.match(r'C3[89][0123489]', place)
    peri_yes = re.match(r'C48[28]', place)
    ovi_yes = re.match(r'C5[4567]', place)
    test_yes = re.match(r'C6[23]', place)
    if pleura_yes:
      place = "pleura"
    elif pleura_yes:
      place = "pleura"
    elif peri_yes:
      place = "peritoneum"
    elif ovi_yes:
      place = "ovary"
    elif test_yes:
      place = "testis"
```

```
    else:
      continue
    dxp = "mesothelioma_" + place
    if not dxhash.has_key(dxp):
      dxhash[dxp] = "0 0 0 0 0 0 0 0 0 0 0 0 0 0 0 0 0 0 0 0"
    age_at_dx = line[24:27]
    age_at_dx = int(age_at_dx)
    if (age_at_dx > 95):
        age_at_dx = 95
    age_at_dx = int(float(age_at_dx) / 5)
    agearray = dxhash[dxp].split(" ")
    agearray[age_at_dx] = str(int(agearray[age_at_dx]) + 1)
    dxhash[dxp] = " ".join(agearray)
keylist = dxhash.keys()
for item in keylist:
    print item + "|" + dxhash[item]
print "The total number of mesotheliomas is " + str(count)
exit
```

Ruby Script

```
#!/usr/local/bin/ruby
require 'mathn'
count = 0
dxhash = {}
filelist = Dir.glob("c:/big/seer2006/*.TXT")
filelist.each do
  |filepathname|
  seer_file = File.open(filepathname, "r")
  seer_file.each do
    |line|
    code1 = line.slice(47,5)
    code2 = line.slice(52,5)
    next unless (code1 =~ /^905/ || code2 =~ /^905/)
    count = count + 1
    place = line.slice(42,4)
    if place =~ /C34[12389]/
      place = "pleura"
    elsif place =~ /C3[89][0123489]/
      place = "pleura"
    elsif place =~ /C48[28]/
      place = "peritoneum"
    elsif place =~ /C5[4567]/
      place = "ovary"
    elsif place =~ /C6[23]/
      place = "testis"
    else
      next
    end
    dxp = "mesothelioma_" + place
```

```
    unless dxhash.has_key? dxp
      dxhash[dxp] = "0 0 0 0 0 0 0 0 0 0 0 0 0 0 0 0 0 0 0 0"
    end
    age_at_dx = line.slice(24,3)
    next if age_at_dx !~ /[01][0-9]{2}/
    age_at_dx = age_at_dx.to_f
    if (age_at_dx > 95)
       age_at_dx = 95
    end
    age_at_dx = (age_at_dx.to_f / 5).to_i
    agearray = dxhash[dxp].split(" ")
    agearray[age_at_dx] = agearray[age_at_dx].to_i
    agearray[age_at_dx] = agearray[age_at_dx] + 1
    dxhash[dxp] = agearray.join(" ")
  end
end
dxhash.each{|key, value| puts key + "|" + value}
puts "The total number of mesotheliomas is " + count.to_s
exit
```

25.2.2 Analysis

The resulting data is shown:

mesothelioma_pleura
0 0 0 3 4 9 27 45 89 189 324 510 691 921 1102 1111 804 325 9 3 12

mesothelioma_peritoneum
0 0 0 2 2 9 9 10 19 32 39 60 88 84 91 63 31 18 3 0

mesothelioma_testis
0 0 0 1 2 0 0 1 2 1 1 9 3 8 5 4 2 2 0 0

mesothelioma_ovary
0 0 0 1 0 0 2 1 3 3 4 2 1 1 1 1 0 1 0 0

Total number of mesotheliomas is 7163

We can see at a glance that there are many more mesotheliomas of the pleura than there are mesotheliomas arising in other sites. The second most common site of mesotheliomas is the peritoneum. Mesotheliomas arising from the testis or the ovary are quite rare.

25.3 Graphic Representation

In the previous section, we produced the distribution of occurrences of mesotheliomas, by age, at different anatomic sites. The output of the data was a list of 20 numbers, corresponding to the 20 age categories ranging from age 0 to age 95 and above. With these data, it is possible to create graphic representations of the data that can be visually scanned, to find trends or anomalies of biologic importance that could not be appreciated by inspecting stark lists of numbers.

25.3.1 Script Algorithm

1. Call the external ImageMagick module into your script.
2. Create an image object.
3. Read an external image into your new image object. In this case, the external image serves as a blank object upon which you will draw the graphic representations of the age distributions of occurrences of mesotheliomas, at four anatomic sites.

   We will use the image file, empty.gif, created as a 520 by 520 pixel gif file available at

   http://www.julesberman.info/book/empty.gif

   Alternately, you can draw your own empty gif file, or you can simply create an empty file, in your script, using ImageMagick.
4. Create a dictionary object to hold the four distributions created in the previous section of this chapter. The keys will be the anatomic sites, and the values will be a string with 18 values corresponding to the number of occurrences of mesotheliomas, for the age range (beginning with age 0 and ending with age 85 and above) collected in the SEER data records.
5. Parse through each key–value pair in the dictionary object.
6. As each key–value pair is parsed, split the value string into an array of 18 items containing the cancer occurrences, by age.
7. Determine the largest number in each array, and save it as a variable.
8. Print the key (the anatomic site where the mesotheliomas arose) at a position on the image object to the left of the position where the graph will appear.
9. Build a graph of the age distribution, using the fraction of the size of each member of the distribution, compared with the biggest item of the distribution multiplied by 50 as the height of each bar. This method guarantees that all four graphs will fit on the image (because no graph will have a height exceeding 50 pixels, and four 50-pixel graphs can easily fit on a canvas that is 520 pixels high). Likewise, each graph will have the same vertical height (50 pixels), regardless of the size of individual members of the distribution. For each member of the distribution, stagger its horizontal location on the graph in 15-pixel increments, beginning with a pixel location 100 pixels from the left border of the image.
10. Step 9 is repeated for each of the four distributions, raising the x-axis by 120 pixels for each succeeding graph.
11. After all the distributions are parsed, write the image object to a newly created image file.

Perl Script

```perl
#!/usr/local/bin/perl
use Image::Magick;
my $image = Image::Magick->new;
$image -> ReadImage("c\:\\ftp\\metajpg\\empty.gif");
$dict{"meso_pleura"} = "0 0 0 3 4 9 27 45 89 189 324 510 691 921
1102 1111 804 430";
$dict{"meso_peritoneum"} = "0 0 0 2 2 9 9 10 19 32 39 60 88 84 91
63 31 21";
$dict{"meso_testis"} = "0 0 0 1 2 0 0 1 2 1 1 9 3 8 5 4 2 2";
$dict{"meso_ovary"} = "0 0 0 1 0 0 2 1 3 3 4 2 1 1 1 1 0 1";
$base_y = 580;
while (($key, $value) = each(%dict))
   {
   @data_array = split(/ /,$value);
   $big_item = 0;
   foreach $item (@data_array)
      {
      if ($item > $big_item)
         {
         $big_item = $item;
         }
      }
   $base_x = 100;
   $base_y = $base_y - 120;
   $image->Annotate(text => $kcy, x => ($base_x - 70), y =>
$base_y);
   foreach $item (@data_array)
      {
      $item = int(($item / $big_item)*50);
      $base_x = $base_x + 15;
      $peak_y = $base_y - $item;
      $image -> Draw (stroke => "black", width => "2", primitive =>
"line",
      points => "$base_x,$base_y $base_x,$peak_y");
      }
   }
$image -> write ("meso.gif");
exit;
```

Python Script

```python
#!/usr/local/bin/python
import Image, ImageDraw
im = Image.open("c:/ftp/metajpg/empty.gif")
draw = ImageDraw.Draw(im)
dict = {}
dict["meso_pleura"] = "0 0 0 3 4 9 27 45 89 189 324 510 691 921
1102 1111 804 430"
```

```
dict["meso_peritoneum"] = "0 0 0 2 2 9 9 10 19 32 39 60 88 84 91 63
31 21"
dict["meso_testis"] = "0 0 0 1 2 0 0 1 2 1 1 9 3 8 5 4 2 2"
dict["meso_ovary"] = "0 0 0 1 0 0 2 1 3 3 4 2 1 1 1 1 0 1"
base_y = 580
for key, value in dict.iteritems():
  data_array = value.split(" ")
  big_item = 0
  for item in data_array:
     item = int(item)
      if item > big_item:
        big_item = item
  base_x = 30
  base_y = base_y - 120
  draw.text((base_x,base_y), key)
  for item in data_array:
    item = int((float(item) / big_item)*50)
    draw.line((base_x,base_y) + (base_x, (base_y - item)), width=1,
fill=000)
    base_x = base_x + 25
im.save("meso.jpg")
exit
```

Ruby Script

```
#!/usr/local/bin/ruby
require 'RMagick'
include Magick
img = Magick::ImageList.new("c:/ftp/metajpg/empty.gif")
gc = Magick::Draw.new
gc.fill_opacity(0)
gc.stroke('black').stroke_width(1)
dict = {}
dict["meso_pleura"] = "0 0 0 3 4 9 27 45 89 189 324 510 691 921
1102 1111 804 430"
dict["meso_peritoneum"] = "0 0 0 2 2 9 9 10 19 32 39 60 88 84 91 63
31 21"
dict["meso_testis"] = "0 0 0 1 2 0 0 1 2 1 1 9 3 8 5 4 2 2"
dict["meso_ovary"] = "0 0 0 1 0 0 2 1 3 3 4 2 1 1 1 1 0 1"
base_y = 580
dict.each do
  |key, value|
  data_array = value.split(/ /)
  base_x = 80
  base_y = base_y - 120
  gc.font_weight("normal")
  gc.text((base_x - 70),base_y, key)
  data_array.each do
    |item|
    item = (((item.to_f / data_array.max.to_f)*50).ceil)
```

```
        gc.line(base_x,(base_y - item), base_x, base_y)
        base_x = 20 + base_x
    end
end
gc.draw(img)
img.write("meso.gif")
exit
```

25.3.2 Analysis

From the numbers, it is clear that pleural mesothelioma occurs much more frequently than peritoneal mesothelioma. Examination of the graphic representations of the data allows us to draw some additional conclusions (Figure 25.3).

There seems to be about a 10-year difference in the peak age of occurrence of pleural and peritoneal mesotheliomas. The peak age of occurrence of pleural mesothelioma is 75 years, and that of peritoneal mesotheliomas is 65 years. Unlike either peritoneal or pleural mesotheliomas, both ovarian and testicular mesotheliomas occur in a younger population than mesotheliomas that occur in the pleura and peritoneum. The peak age

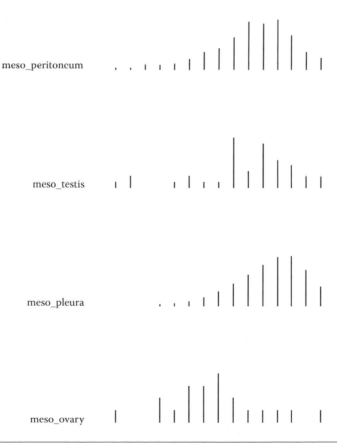

Figure 25.3 Age-incidence curves for mesotheliomas occurring at four different anatomic locations. The curves do not have the same shape or the same age locations of incidence peaks.

of occurrence is of ovarian mesothelioma is much lower than mesotheliomas occurring in pleura or peritoneum.

In Chapter 26, we will discuss the biological relevance of findings based on age distributions of the occurrence of tumors. For now, it suffices to say that the SEER data sets permit us to evaluate how the incidence of tumors may vary with a variety of factors included in the SEER records. Simple modifications of the methods provided in this chapter allow us to examine raw data and graphic representations of tumor occurrences organized by age, ethnicity, race, gender, time, anatomic site, tumor subtype (recognized variants of a tumor that are recorded by ICD code), tumor type (name of tumor), and tumor class (biologic category into which one or more tumors may fall).

Exercises

1. Using Perl, Python, or Ruby, modify the script to provide the total number of mesotheliomas at each anatomic site for which there is an ICD topography code and for which there is a record containing a mesothelioma diagnosis.
2. Using Perl, Python, or Ruby, modify the script to find the age distributions for mesotheliomas at the different sites, stratified by gender.
3. Using Perl, Python, or Ruby, modify the script to find the age distributions for mesotheliomas aggregated for all sites, and separated by race/ethnicity (white persons, black persons, Hispanic persons, and Asian persons).
4. Using Perl, Python, or Ruby, for each type of mesothelioma, write a script that determines the total number of mesotheliomas, of the type, in the SEER data sets, and the average age of occurrence of each type. The four types of mesothelioma are

   90503 = mesothelioma (general designation for mesotheliomas, without subtyping)
   90513 = fibrous mesothelioma
   90523 = epithelial mesothelioma
   90533 = biphasic mesothelioma

5. Using Perl, Python, or Ruby, modify the script to determine the age distributions of some tumor other than mesothelioma. Does this tumor also have the same discordance of curve shape and peak incidence that we have seen in mesothelioma?

26

CASE STUDY

Bimodal Tumors

Most tumors have a simple, smooth age distribution, with a single peak (Figure 26.1). Not all cancers have a single age-of-occurrence peak. Some have two or more peaks of occurrence.

Well-known examples of recognized bimodal cancers are Hodgkin lymphoma (which has two peaks in occurrence: in young adults and in middle-aged adults), and Kaposi's sarcoma (which has two peaks in occurrence: in young people, with AIDS, and in older men, unassociated with AIDS).

Here are the causes for cancer multimodality (multiple peaks in a graph of cancer occurrences by age).

1. Two different types of cancer, occurring in two different peak age groups, are mistakenly assigned the same name.
2. Two or more environmental causes for the same cancer target different ages.
3. Two or more genetic causes for the same cancer have different latencies (time after birth when the cancer become clinically detected).
4. Two or more human subpopulations, with different susceptibilities to developing the cancer, are pooled in the sample population.
5. Faulty or insufficient data. Bimodality may be a distortion due to poor data that does not adequately conform to the naturally occurring (unimodal) distribution.
6. False conclusion based on accurate data. Statistical analysis is a form of intelligent guessing. Scientists must never assume that any of their statistical conclusions are correct. All conclusions must be constantly reexamined in light of new findings.
7. Combinations of the above.

We see multimodal distributions when we mistakenly call several different kinds of cancer by the same name. For example, lung cancer in young persons may have a specific mutation that distinguishes these cancers from lung cancers occurring in an older population. In particular, midline carcinoma of children and young adults has a characteristic gene arrangement involving the NUT gene (lacking in lung cancers occurring in older adults). Lung cancer of the young is grouped with lung cancer of older adults. However, they are different cancers, with different tumor genetics. It may turn out that lung cancer of the young may respond to a different treatment than lung cancers caused by smoking.

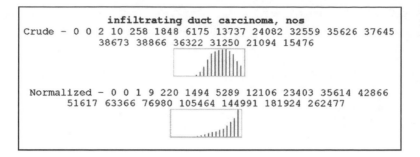

Figure 26.1   A unimodal peak for infiltrating ductal carcinoma of the breast.

Finally, we must consider that it is possible that some multimodal curves are simply an artifact produced by the way we collect and analyze data. If the pathologists who rendered the diagnoses used in the SEER data set were wrong (i.e., rendered misdiagnoses), we would expect multimodality on that basis (representing the different tumors included under a category that should have included only one kind of cancer).

This actually happens. A good example is malignant fibrous histiocytoma. Current thinking is that this diagnostic entity has been used as a grab-bag diagnosis for sarcomas that do not fit well into any particular category. There is substantial evidence that many cases of malignant fibrous histiocytoma would have been better diagnosed as leiomyosarcomas or liposarcomas or fibrosarcomas, and a host of rare sarcomas, each with its own characteristic age distribution. By blending these different tumors under a single name, you also blend the age distributions of the reported population.

The shape of the curve of cancer occurrences, by age, for the different types of cancer, is a fascinating area of research. If we understood why some cancer curves are bimodal, we could enhance our knowledge of carcinogenesis (the developmental process of cancer) and tumor diagnosis (the features that identify a cancer and that separate a particular type of cancer from all other types of cancer). We could also learn a lot about the meaning of the data that we collect on cancers, and the ways that this data can be analyzed. Most importantly, the insights gained can save lives, by uncovering preventable cancers, and by finding new classes and subclasses of cancer that may benefit from innovative cancer treatments.

In this chapter, we will make a scientific contribution to the field of multimodal cancer incidence, by age, by finding every cancer in the SEER database, with a multimodal distribution. The project consists of three steps:

1. Determine the age distribution of every type of cancer included in the SEER data sets, covering the years 1973–2006 (about 650 diagnostic types).
2. Represent each distribution as a graph.
3. Display all 650 graphs within a single document that can be visually scanned, to find distributions that are bimodal (or multimodal).
4. Collect the bimodal graphs in a single document that can be shared with the scientific community.

Our first task is to prepare the age distribution of every type of cancer included in the SEER data sets. In Chapter 7, we learned how to collect cancers by diagnostic code from the SEER data sets, and to determine the distributions of occurrences of a cancer by age. In this chapter, we will take this process two steps further, as we determine the rate of occurrence of these cancers, as a fraction of the total population of the people in the age group, in the United States, using U.S. census data, and applying the analysis to every type of cancer contained in the SEER data sets. The script is similar to the script that produced the age distributions of mesotheliomas at different anatomic sites (Chapter 25). The key difference in this script is that the age-distributions for the different cancers will be expressed as a population rate, not simply as raw occurrence numbers. To produce a population rate, we need to know the number of people, in the U.S. population, in each age group.

The U.S. Census Bureau provides a simple file containing the U.S. population, for each year of age, from 0 to 100+ (Figure 26.2).

In this figure, the first item is the date, the second is the age, the third is the total population for the age, the fourth is the male population for the age, and the fifth is the female population for the age.

The file can be downloaded from

http://www.census.gov/popest/archives/EST90INTERCENSAL/
US-EST90INT-07/US-EST90INT-07-2000.csv

or from

http://www.julesberman.info/book/censuage.txt

The number of occurrences of any cancer, in an age group, are divided by the number of people, in the United States, in the age group, and then multiplied by 100,000. This yields an incidence rate.

26.1 Script Algorithm

1. Open the external file, censuage.txt.
2. Parse through every line of the censuage.txt file.
3. The censuage.txt file provides population numbers, for the United States for ages of individual years, up to age 100. We want to bin these ages into 20 five-year intervals (i.e., 0 up to five, 5 up to 10, 10 up to 15, etc.) The first item (following the 0th item) in the census record is the age, followed by the second item, the U.S. population for the age. We divide each age by 5, round to an integer, and sum all of the population values for the rounded integer, to build a dictionary object whose keys are the 20 age groups, and whose values are the aggregate population of the age group.

```
"January 1 2000","66",1861037,863707,997330
"January 1 2000","67",1914140,881318,1032822
"January 1 2000","68",1866063,856310,1009753
"January 1 2000","69",1904863,867606,1037257
"January 1 2000","70",1871126,840143,1030983
"January 1 2000","71",1798289,802794,995495
"January 1 2000","72",1785616,786484,999132
"January 1 2000","73",1717861,748131,969730
"January 1 2000","74",1690634,722627,968007
"January 1 2000","75",1650457,695132,955325
"January 1 2000","76",1550745,645326,905419
"January 1 2000","77",1473426,605532,867894
"January 1 2000","78",1427387,577777,849610
"January 1 2000","79",1301829,515374,786455
"January 1 2000","80",1197861,462953,734908
"January 1 2000","81",1066033,403284,662749
"January 1 2000","82",979704,364550,615154
"January 1 2000","83",880036,315548,564488
"January 1 2000","84",802873,279166,523707
"January 1 2000","85",726094,242858,483236
"January 1 2000","86",632874,203817,429057
"January 1 2000","87",550951,171556,379395
"January 1 2000","88",464939,138564,326375
"January 1 2000","89",400877,113529,287348
"January 1 2000","90",325719,88900,236819
"January 1 2000","91",266325,68862,197463
"January 1 2000","92",216653,54020,162633
"January 1 2000","93",168821,39564,129257
"January 1 2000","94",132885,29948,102937
"January 1 2000","95",97375,20930,76445
"January 1 2000","96",72361,14621,57740
"January 1 2000","97",52195,10201,41994
"January 1 2000","98",36892,6957,29935
"January 1 2000","99",26984,5148,21836
"January 1 2000","100",49940,9863,40077
```

Figure 26.2 Population of the United States, January 1, 2000, stratified by age and gender. The first item is the date, the second is the age, the third is the total population for the age, the fourth is the male population for the age, and the fifth is the female population for the age.

4. Use the icdo3.txt file (see appendix), containing the list of neoplasm names and codes used in the SEER data files, to create a dictionary object wherein the keys are code numbers and the values are the corresponding neoplasm names.

5. Parse through SEER files, line by line. Each line of a SEER file is the record of a cancer occurrence, and there are over 3.7 million lines that will be parsed. Instructions for obtaining the free, public use SEER files are found in the appendix. In this example script, the SEER files are found in the \seer2006 subdirectory.

6. As each line of the file is parsed, extract the 5-character substring that begins at character 48 and the 5-character substring that begins at character 53. These represent the ICD-O code for the record. The string beginning at character 53 is the code for the newer version of ICD-O. If this string has a code that is

contained in the version of ICD-O that we are using (in the icdo3.txt file), we will use this code, rather than the code contained in the substring that begins at character 48.

7. As each line of the file is parsed, extract the three-digit number representing the age of the patient at the time that the tumor was diagnosed. The age is found in record bytes 25, 26, and 27.

8. Bin each age into one of 20 bins by dividing the age by 5, taking the integer value of the result, and lumping all ages 95 and above to the same bin (the 20th bin).

9. For each record, increment (by 1) the number of occurrences of the diagnosis (for the record), in the bin corresponding to the patient age listed for the record, and put the record diagnosis and the incremented age distribution for the diagnosis, as a key–value pair for a dictionary object.

10. After all of the files are parsed, parse through the dictionary object containing all of the diagnostic code/age distributions.

11. For each encountered value (a string containing 20 numbers corresponding to the number of occurrences of the tumor in each of the 20 age groups), normalize the occurrences into a population rate, using the following formula:

age group rate = crude number of occurrences in the age divided by the U.S. population for the age group multiplied by 10000000

12. Print out the keys and values into two separate output files: seer_all.txt and seer_adj.txt.

13. To the seer_all.txt file, for each of the cancers in the dictionary object, print the name of the cancer (the value of the ICD-O code contained in the dictionary object created by parsing through the icdo3.txt file), followed by the age distributions of the crude cancer occurrence numbers for each age group, followed by the population incidence rate normalized against the U.S. population for the age group.

14. To the seer_adj.txt file, print the name of the cancer (the value of the ICD-O code contained in the dictionary object created by parsing through the icdo3.txt file), followed by the population incidence rate normalized against the U.S. population for the age group.

Perl Script

```
#!/usr/local/bin/perl
open (ICD, "c\:\\ftp\\censuage\.txt")||die"cannot";
$line = " ";
while ($line ne "")
  {
  $line = <ICD>;
  @linearray = split(/\,/, $line);
    {
```

```
      $age_at_dx = $linearray[1];
      $age_at_dx =~ s/\"//g;
      if ($age_at_dx > 95)
          {
          $age_at_dx = 95;
          }
      $age_at_dx = int($age_at_dx / 5);
      $total{$age_at_dx} = $total{$age_at_dx} + $linearray[2];
      }
  }
close ICD;
open (ICD, "c\:\\ftp\\icdo3\.txt");
$line = " ";
while ($line ne "")
  {
  $line = <ICD>;
  if ($line =~ /([0-9]{4})\/([0-9]{1}) +/o)
    {
    $code = $1 . $2;
    $term = $';
    $term =~ s/ *\n//o;
    $term = lc($term);
    $dictionary{$code} = $term;
    }
  }
close ICD;
opendir(SEERDIR, "c\:\\big\\seer2006") || die ("Unable to open
directory");
@files = readdir(SEERDIR);
closedir(SEERDIR);
open (ALLOUT, ">c\:\\ftp\\seer_all.txt");
open (ADJOUT, ">c\:\\ftp\\seer_adj.txt");
chdir("c\:\\big\\seer2006");
foreach $datafile (@files)
  {
  next if ($datafile !~ /.txt/i);
  open (TEXT, $datafile);
  $line = " ";
  while ($line ne "")
    {
    $line = <TEXT>;
    $dx = substr($line, 47, 5);
    $dx2 = substr($line, 52, 5);
    if (exists($dictionary{$dx2}))
        {
        $dx = $dx2;
        }
    unless (exists($dxhash{$dx}))
```

```
      {
      $dxhash{$dx} = "0 0 0 0 0 0 0 0 0 0 0 0 0 0 0 0 0 0 0 0";
      }
    $age_at_dx = substr($line,24,3);
    if ($age_at_dx > 95)
        {
        $age_at_dx = 95;
        }
    $age_at_dx = int($age_at_dx / 5);
    @agearray = split(" ", $dxhash{$dx});
    $agearray[$age_at_dx]++;
    $dxhash{$dx} = join(" ", @agearray);
    }
  close TEXT;
  }
while (($key, $value) = each(%dxhash))
  {
  if (exists($dictionary{$key}))
      {
      @value_array = split(/ /, $value);
      for ($i=0;$i<20;$i++)
          {
          $value_array[$i] = int(($value_array[$i] / $total{$i})
*10000000);
          }
      $rate_value = join(" ", @value_array);
      print ALLOUT "$dictionary{$key}\n$value\n$rate_value\n\n";
      print ADJOUT "$dictionary{$key}\|$rate_value\n";
      }
  }
exit;
```

Python Script

```
#!/usr/local/bin/python
import os, re, string
census_in = open("c:/ftp/censuage.txt", "r")
total = {}
linearray = []
for line in census_in:
  linearray = line.split(",")
  age_at_dx = linearray[1]
  age_at_dx = re.sub(r'"', '', age_at_dx)
  age_at_dx = float(age_at_dx)
  if age_at_dx > 95:
    age_at_dx = 95
  age_at_dx = int(age_at_dx/5)
  if total.has_key(age_at_dx):
    total[age_at_dx] = total[age_at_dx] + int(linearray[2])
```

```
    else:
      total[age_at_dx] = int(linearray[2])
census_in.close()
f = open("c:\\ftp\\icdo3.txt", "r")
codehash = {}
subhash = {}
agearray = []
for line in f:
  linematch = re.search(r'([0-9]{4})\/([0-9]{1}) +(.+)$', line)
  if (linematch):
    icdcode = linematch.group(1) + linematch.group(2)
    term = string.rstrip(linematch.group(3))
    codehash[icdcode] = term
f.close()
filelist = os.listdir("c:/big/seer2006")
os.chdir("c:/big/seer2006")
for file in filelist:
  seer_file = open(file, "r")
  for line in seer_file:
    code1 = line[47:52]
    code2 = line[52:57]
    if codehash.has_key(code2):
      code1 = code2
    age_at_dx = int(line[24:27])
    if (age_at_dx > 95):
      age_at_dx = 95
    age_at_dx = int(float(age_at_dx)/5)
    if not subhash.has_key(code1):
      subhash[code1] = "0 0 0 0 0 0 0 0 0 0 0 0 0 0 0 0 0 0 0 0"
    agearray = subhash[code1].split(" ")
    old_value = agearray[age_at_dx]
    new_value = int(old_value) + 1
    agearray[age_at_dx] = str(new_value)
    subhash[code1] = " ".join(agearray)
all_out = open("c:/ftp/seer_all.txt", "w")
adj_out = open("c:/ftp/seer_adj.txt", "w")
for key,value in subhash.iteritems():
  if codehash.has_key(key):
    value_array = value.split(" ")
    for i in range(len(value_array)):
      value_array[i] = str(int(float(value_array[i]) /
float(total[i]) * 10000000))
    rate_value = " ".join(value_array)
    print>>all_out, codehash[key] + "\n" + value + "\n" +
rate_value + "\n"
    print>>adj_out, codehash[key] + "|" + rate_value
exit
```

Ruby Script

```
#!/usr/local/bin/ruby
require 'mathn'
census_in = File.open("c:/ftp/censuage.txt", "r")
total = Hash.new(0)
dxhash = {}
linearray = []
total = Hash.new(0)
census_in.each do
  |line|
  linearray = line.split(",")
  age_at_dx = linearray[1]
  age_at_dx = age_at_dx.gsub(/\"/, "").to_i
  if age_at_dx > 95
    age_at_dx = 95;
  end
  age_at_dx = (age_at_dx.to_f / 5).to_i
  total[age_at_dx] = total[age_at_dx] + linearray[2].to_i
end
census_in.close
icd_in = File.open("c:/ftp/icdo3.txt", "r")
dictionary = {}
icd_in.each_line do
  |line|
  if line =~ /([0-9]{4})\/([0-9]{1}) +/
    code = $1 + $2
    term = $'
    term = term.sub(/ *\n/, "") if term =~ / *\n/
    dictionary[code] = term
  end
end
icd_in.close
all_out = File.open("seer_all.txt", "w")
adj_out = File.open("seer_adj.txt", "w")
filelist = Dir.glob("c:/big/seer2006/*.TXT")
filelist.each do
  |filepathname|
  seer_file = File.open(filepathname, "r")
  seer_file.each do
    |line|
    code1 = line.slice(47,5)
    code2 = line.slice(52,5)
    code1 = code2 if dictionary.has_key? code2
    unless dxhash.has_key? code1
      dxhash[code1] = "0 0 0 0 0 0 0 0 0 0 0 0 0 0 0 0 0 0 0 0 0"
    end
    age_at_dx = line.slice(24,3)
```

```
     next if age_at_dx !~ /[01][0-9]{2}/
     age_at_dx = age_at_dx.to_f
     if (age_at_dx > 95)
        age_at_dx = 95
     end
     age_at_dx = (age_at_dx.to_f / 5).to_i
     agearray = dxhash[code1].split(" ")
     agearray[age_at_dx] = agearray[age_at_dx].to_i
     agearray[age_at_dx] = agearray[age_at_dx] + 1
     dxhash[code1] = agearray.join(" ")
  end
end
dxhash.each do
  |key, value|
  if dictionary.has_key?(key)
     value_array = value.split(" ")
     value_array.each_index do
       |i|
       value_array[i] = ((value_array[i].to_f / total[i].to_f) *
10000000).to_i
     end
     rate_value = value_array.join(" ")
     all_out.puts dictionary[key] + "\n" + value + "\n" +
rate_value + "\n"
     adj_out.puts dictionary[key] + "|" + rate_value
  end
end
exit;
```

26.2 Analysis

The age distributions of 650 cancers in the SEER database are collected in two output files (seer_all.txt and seer_adj.txt).

We have already seen, in Chapters 3 and 25, how we can represent one or many age distributions, as simple graphs. We do not need to repeat the code examples from these prior chapters. Some of the graphic representations of the 650 cancers, with age-of-occurrence rate distributions, are shown in Figure 26.4.

A file of the 650 cancer age distributions is available for download at

http://www.julesberman.info/seerdist.pdf

We can visually inspect all of these cancers, looking for multimodal distributions. Several dozen of the 650 cancers from the SEER data sets were collected. The multimodal graphs were assembled into a single file that is available at

http://www.julesberman.info/bimode.pdf

```
acute promyelocytic leuk.,t(15;17)(q22;q11-12)
22  23  41  52  71  62  110  88  97  94  97  72  88  69  87  83  34  23  6  3
11  11  20  25  38  32  54  38  43  47  55  53  81  72  98  112  69  82  54  89


ependymoma, anaplastic
46  18  13  6  4  13  3  9  12  7  7  6  3  3  1  0  0  1  0  0
24  8  6  2  2  6  1  3  5  3  4  4  2  3  1  0  0  3  0  0


giant cell and spindle cell carcinoma
0  0  0  1  0  1  2  8  5  18  28  28  31  30  38  24  18  14  3  1
0  0  0  0  0  0  0  3  2  9  16  20  28  31  42  32  36  50  27  29


paget disease and intraductal ca.
0  0  0  0  0  3  26  55  76  109  113  139  169  176  165  169  108  68  15  7
0  0  0  0  0  1  12  24  34  54  65  104  157  184  186  228  219  244  135  208


angiomatous meningioma
0  1  1  0  0  1  1  0  3  3  7  9  4  6  3  1  2  0  1  0
0  0  0  0  0  0  0  0  1  1  4  6  3  6  3  1  4  0  9  0


pituitary carcinoma, nos
0  0  0  0  0  0  0  0  1  1  1  1  0  2  0  0  0  0  0  0
0  0  0  0  0  0  0  0  0  0  0  2  0  0  0  0  0  0  0  0


chronic myeloid leukemia, nos
51  30  56  116  195  325  449  556  562  606  721  789  905  1052  1093  1150  982  608  247  69
26  14  27  57  105  168  221  245  252  303  415  590  841  1102  1233  1553  1993  2190  2224  2055


bronch.-alv. carc., mixed mucin. and non-mucinous
0  0  0  0  0  1  0  0  0  1  3  4  4  2  5  8  2  3  0  0
0  0  0  0  0  0  0  0  0  0  1  2  3  2  5  10  4  10  0  0
```

Figure 26.3 Several of the 650 cancer age distributions from the SEER_all.txt file. Each cancer entry consists of the name of each cancer, followed by the crude occurrences of the cancer by age group, followed by the occurrence rates of the cancer by age group.

Figure 26.5 is a sample of a bimodal graph, for serous papillary cystic tumor of borderline malignancy. Two peaks are clearly evident. Here is another bimodal graph; this time for rhabdomyosarcoma (Figure 26.6). The curves for the crude and the normalized age distributions of rhabdomyosarcoma look very different from one another. When we normalize age distributions against a standard population, the two areas of the graph that look most different are the youngest age groups and the oldest age groups. This is because the U.S. population of children is very large. A few occurrences of cancer, among children, produce a small blip in the age-normalized graph because the rate of occurrence of a few tumors in a large population is small. At the other extreme of the graph, people aged 85 and higher, the effect is the opposite. The population of people over the age of 85 is small, so a few occurrences of cancer in this population can produce a large age-specific rate of cancer. A cancer that occurs preferentially in the very young and the very old (such as rhabdomyosarcoma), will produce starkly different graphs when we compare crude occurrences with normalized rates distributed over a range of ages. Nonetheless, the basic bimodality of the rhabdomyosarcoma curves is seen in both varieties of graphs.

We have seen that serous papillary cystic tumor of borderline malignancy has a bimodal distribution. Is bimodality a feature of all histologic subtypes of borderline malignancy? Apparently, yes. We see bimodality in the graph of mucinous cystic tumor of borderline malignancy. Papillary cystadenoma of borderline malignancy may have two peaks. Two peaks are not quite so obviously present as the distribution of mucinous cystic tumor of borderline malignancy (Figure 26.7).

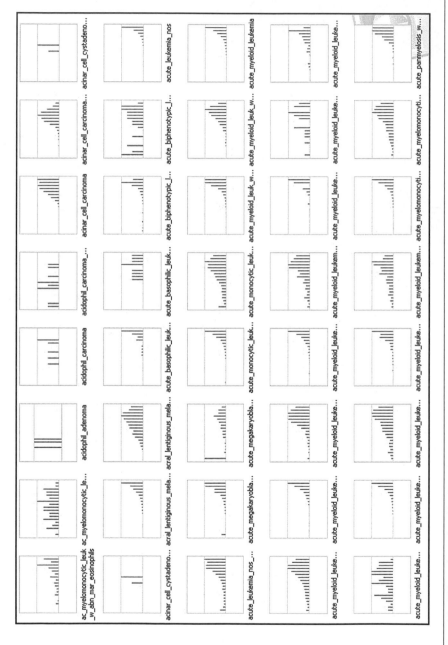

Figure 26.4   A few of the 650 graphs produced by the script.

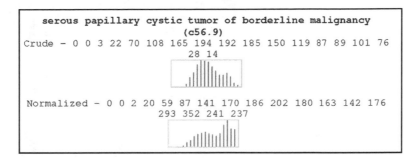

Figure 26.5 Age distribution of occurrences of serous papillary cystic tumor of borderline malignancy. The top graph represents the crude occurrences by age, in the SEER data sets. The bottom graph represents the occurrence rates, normalized against the population of the age group, as determined by the U.S. Census, for the year 2000.

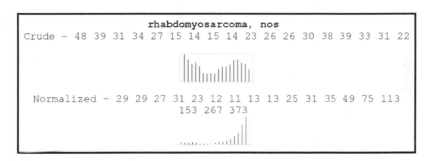

Figure 26.6 Age distribution of occurrences of rhabdomyosarcoma. The top graph represents the crude occurrences by age, in the SEER data sets. The bottom graph represents the occurrence rates, normalized against the population of the age group, as determined by the U.S. Census, for the year 2000.

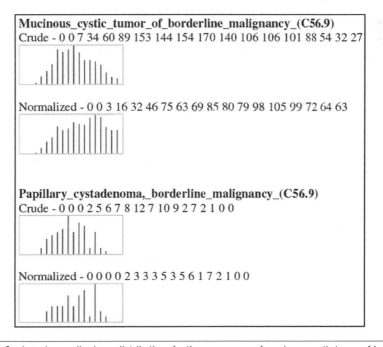

Figure 26.7 Crude and normalized age distributions for the occurrences of mucinous cystic tumor of borderline malignancy and papillary cystadenoma of borderline malignancy.

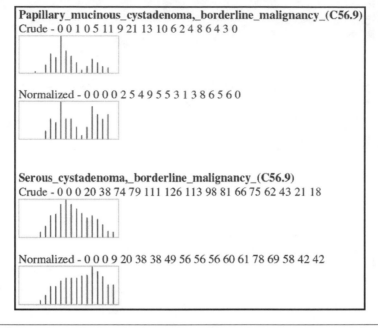

Figure 26.8 Crude and normalized age distributions for papillary mucinous cystadenoma borderline malignancy and serous cystadenoma borderline malignancy.

When we look at papillary mucinous cystadenoma borderline malignancy, the bimodality of the graphs is obvious. The graphs of serous cystadenoma borderline malignancy also shows two peaks, though the second peak (in the older population) is small (Figure 26.8).

It is possible to see multiple peaks when none occur. Elise Whitley and Jonathan Ball (in "Statistics review 1: Presenting and summarising data" *Critical Care* 6:66–71, 2002) explained that multiple peaks can come from data rounding errors (up or down). When you are inspecting graphs, and peaks occur at values ending with zero (50, 100, 200, etc.), you should always suspect that the data artifactually aggregates at values to which your measurements have been rounded.

You can also miss multimodality, even when it is present. Schilling and co-workers have shown that bimodality can be missed if the difference in peak locations is small relative to the standard deviations of the populations. For example, a mixture of two normal distributions with equal standard deviations is bimodal only if the two means differ by at least twice the common standard deviation (Schilling M. F., Watkins A. E., Watkins W. Is human height bimodal? *Am. Statistician* 56:223–229, 2002).

Still, there are many reasons to believe that many of the bimodal distributions found in the SEER data sets reveal true biological features of the cancer populations.

1. The multimodal peaks are rare among cancers. Of the more than 650 cancers collected in the complete file of cancer occurrences by age, only a couple dozen show multimodality. If multimodality were a systemic artifact, would you not expect to see it occurring in the majority of cancer distributions?

2. The SEER data reproduces multimodal peaks in the same cancers for which multimodal peaks have been established from other data sources. For example, the SEER data shows bimodal peaks for Hodgkin lymphoma, Kaposi sarcoma, and secretory carcinoma of the breast.

3. The SEER data provides very large numbers of cases for many of the cancers for which bimodal peaks are found. The shape of the curves cannot be attributed to sparse data, in these cases.

4. We will also see that there is internal consistency of the observation of multimodality within the SEER data. In some cases, data is collected within SEER on a single tumor and its variant forms. For example, the borderline tumors of the ovary are listed under several closely related terms. In the case of borderline ovarian tumors and its variants, multimodality was preserved among the related cancers.

5. Multimodality is a phenomenon that we would expect to occur in human cancers, because we know that a given type of cancer can have many different causes, and that these causes may exert biological effects in persons of specific age groups.

When we examine the distribution files, we can discover new hypotheses about neoplasms in general, and bimodal tumors, in particular. The persistent message from this chapter is that multimodality in a disease distribution is a puzzle worth investigating.

Exercises

1. Using Perl, Python, or Ruby, write a script that takes the 650 tumor-specific age distributions, produced by this chapter's script, to produce 650 images, each containing two vertically aligned graphs, one displaying the occurrences of tumors by age group, and the other displaying the rate of occurrence of tumors by age group.

2. Using Perl, Python, or Ruby, modify your script from Exercise 1 to include an embedded header, for each of the 650 images, containing your name, the date of creation of the image, and the name of the tumor whose age distributions are featured in the image.

3. Using Perl, Python, or Ruby, modify your script from Exercise 2 to produce a Web page that displays all of the 650 images produced in Exercises 1 and 2.

4. In Chapter 25, we found the age distributions of the occurrences of mesotheliomas in four different anatomic sites. Using Perl, Python, or Ruby, modify the script from Chapter 25 to provide a distribution of the rates of occurrence of mesothelioma for the same age intervals and for the same anatomic sites.

27

CASE STUDY

The Age of Occurrence of Precancers

I recently wrote a medical book, entitled *Precancer: The Beginning and the End of Cancer* (Jones and Bartlett Publishers, 2010). This book was devoted to a specific type of lesion pathologists encounter that precedes the development of cancer, sometimes by many years. Unlike cancers, which grow relentlessly, invade neighboring tissues, and often metastasize to distant organs, precancers are usually small noninvasive lesions that can be treated quite easily. When we treat a precancer, we are eradicating the cancer that might have developed from the precancer. In the book, I argue that the prevention and treatment of precancers is the most effective way to reduce the number of cancer-caused deaths.

Because precancers precede the development of cancers, we should be able to demonstrate that, in a large population, the average age of occurrence of a precancer must occur earlier than the average age of occurrence of the cancers that arise from the precancer. Furthermore, once we know the average age of development of a precancer, and we know the average age of development of the subsequent cancer, we can determine that length of time required for a precancer to progress to a cancer.

In this chapter, we will look at the ages of occurrence of precancers and cancers of the uterine cervix, and we will determine the average age of occurrence of all of the types of precancer and cancer at this anatomic location.

27.1 Script Algorithm

1. Open the icdo3.txt file (see appendix), containing the ICD-O (International Classification of Diseases—Oncology) codes and corresponding neoplasm names, used in the SEER data records.
2. Build a dictionary object wherein the keys are the 5-character ICD-O codes and the values are the corresponding neoplasm names.
3. Parse through SEER files (see appendix), line by line. Each line of a SEER file is the record of a cancer occurrence, and there are over 3.7 million lines that will be parsed. Instructions for obtaining the free, public use SEER files is found in the appendix. In this example script, the SEER files are found in the c:\big\seer2006 subdirectory.
4. As each line of the file is parsed, extract the 5-character substring that begins at character 48 and the 5-character substring that begins at character 53. These

represent the ICD-O code for the record. The string beginning at character 53 is the code for the newer version of ICD-O. If this string has a code that is contained in the version of ICD-O that we are using (in the icdo3.txt file), we will use this code, rather than the code contained in the substring that begins at character 48.

5. The 4-character sequence for the topographic code for (the anatomic location for the primary growth site of the neoplasm) is contained in bytes 43 to 46 of each SEER record. The appendix has instructions for obtaining the topographic codes used in the SEER records. Codes for the anatomic sites where cervical cancer may arise are the following:

C530 Endocervix
C531 Exocervix
C538 Overlapping lesion of cervix uteri
C539 Cervix uteri

These four codes, and no others, fit a pattern matched by "C53" appearing at the beginning of the record substring that contains the topographic code.

When each SEER record is parsed, extract bytes 43, 44, and 45 of the record (i.e., the 42nd, 43rd, and 44th characters of the record), and match them against "C53". Unless there is a match, return to the beginning of the iteration loop, and get the next line in the file.

6. Create a dictionary object whose keys are the encountered neoplasm codes, and whose values are the incremented tally of the number of encountered SEER records that contain the code.

7. Each SEER tumor record contains the age (in years) of the person, at the time of diagnosis of the tumor. The age is contained in a three-character entry occupying bytes 25 to 27 of the record. Extract the age from the record, and put it into a variable.

8. Create another dictionary object whose keys are the encountered neoplasm codes and whose values are a string consisting of the sum of all of the ages encountered that have the diagnosis referenced by the corresponding key, followed by the word "and", followed by the number of persons encountered that have the diagnosis referenced by the corresponding key. When the entire set of SEER records has been parsed, we will be able to compute the average age at the time of diagnosis, of all persons whose tumors have the same code number.

9. After all of the SEER records have been parsed, iterate through every key–value pair in the dictionary object containing the summed ages and total number of records for each neoplasm occurring in the cervix.

10. Print a formatted output, consisting of the average age of occurrence of each tumor, followed by the number of cases of the tumor, followed by the name of the tumor.

Perl Script

```perl
#!/usr/local/bin/perl
open (ICD, "c\:\\ftp\\icdo3\.txt");
$line = " ";
while ($line ne "")
  {
  $line = <ICD>;
  if ($line =~ /([0-9]{4})\/([0-9]{1}) +/o)
    {
    $code = $1 . $2;
    $term = $';
    $term =~ s/ *\n//o;
    $term = lc($term);
    $dictionary{$code} = $term;
    $agehash{$code} = "0 and 0";
    }
  }
close ICD;
opendir(FTPDIR, "C\:\\BIG\\SEER2006") || die ("Unable to open
directory");
@files = readdir(FTPDIR);
closedir(FTPDIR);
chdir("C\:\\SEER");
foreach $datafile (@files)
  {
  open (TEXT, $datafile);
  $line = " ";
  while ($line ne "")
    {
    $line = <TEXT>;
    $code = substr($line, 47, 5);
    $code2 = substr($line, 52, 5);
    if (exists($dictionary{$code2}))
      {
      $code = $code2;
      }
    next if ((substr($line,42,3) ne "C53"));
    $codecount{$code}++;
    if (exists($dictionary{$code}))
      {
      $age = substr($line, 24, 3);
      $oldcode = $agehash{$code};
      $oldcode =~ / and /;
      $oldage = $`;
      $oldcount = $';
      $newage = $oldage + $age;
      $newcount = $oldcount + 1;
      $agehash{$code} = $newage . " and " . $newcount;
```

```
            }
        }
      }
while((my $key, my $value) = each (%agehash))
    {
    next if ($value eq "0 and 0");
    next if ($codecount{$key} < 80);
    $number = $codecount{$key};
    $number = "000000" . $number;
    $number = substr($number,-6,6);
    $value =~ / and /;
    $totalage = $`;
    $count = $';
    $average = int($totalage / $count);
    $average = "00" . $average;
    $average = substr($average,-3,3);
    push(@outarray, "$average $number $dictionary{$key}");
    }
print join("\n", (sort(@outarray)));
exit;
```

Python Script

```
#!/usr/local/bin/python
import sys, os, re, string
icd_file = open("c:\\ftp\\icdo3.txt", "r")
dictionary = {}
agehash = {}
codecount = {}
outarray = []
for line in icd_file:
    linematch = re.match(r'([0-9]{4})\/([0-9]{1})(.+)', line)
    if linematch:
        code = linematch.group(1) + linematch.group(2)
        term = linematch.group(3)
        term = re.sub(r' *\n', "", term)
        term = term.lower()
        term = term.lstrip()
        dictionary[code] = term
        agehash[code] = "0 and 0"
        codecount[code] = 0
icd_file.close()
filelist = os.listdir("c:\\big\\seer2006")
os.chdir("c:\\big\\seer2006")
for file in filelist:
  seer_file = open(file, "r")
  for line in seer_file:
      code = line[47:52]
      code2 = line[52:57]
      if (dictionary.has_key(code2)):
```

```
            code = code2
         if not line[42:45] == "C53":
            continue
         if dictionary.has_key(code):
            codecount[code] = codecount[code] + 1
            age = int(line[24:27])
            oldcode = agehash[code]
            oldcodematch = re.match(r'(.+) and (.+)', oldcode)
            if oldcodematch:
               oldage = oldcodematch.group(1)
               oldcount = oldcodematch.group(2)
            newage = int(oldage) + age
            newcount = int(oldcount) + 1
            agehash[code] = str(newage) + " and " + str(newcount)
   seer_file.close()
for key, value in agehash.iteritems():
   if (value == "0 and 0"):
      continue
   if (codecount[key] < 80):
      continue
   number = codecount[key]
   number = "000000" + str(number)
   number = number[-6:]
   valuematch = re.match(r'(.+) and (.+)', value)
   if valuematch:
      totalage = int(valuematch.group(1))
      count = int(valuematch.group(2))
   if (count > 0):
      average = int(float(totalage) / count)
      average = str(average)
      outarray.append(average + " " + number + " " + dictionary[key])
outarray = sorted(outarray)
for item in outarray:
   print item
exit
```

Ruby Script

```
#!/usr/local/bin/ruby
require 'mathn'
icd_file = File.open("c:/ftp/icdo3.txt", "r")
dictionary = {}
agehash = {}
codecount = {}
outarray = []
icd_file.each_line do
   |line|
   if (line =~ /([0-9]{4})\/([0-9]{1}) +/o)
      code = $1 + $2
      term = $'
```

```ruby
      term.sub!(/ *\n/, "")
      term = term.downcase
      dictionary[code] = term
      agehash[code] = "0 and 0"
      codecount[code] = 0
    end
  end
icd_file.close
filelist = Dir.glob("c:/big/seer2006/*.TXT")
filelist.each do
  |filepathname|
  seer_file = File.open(filepathname, "r")
  seer_file.each_line do
    |line|
    code = line.slice(47,5)
    code2 = line.slice(52,5)
    if (dictionary.has_key?(code2))
      code = code2
    end
    next unless (line.slice(42,3).eql?("C53"))
    if dictionary.has_key?(code)
      codecount[code] = codecount[code] + 1
      age = line.slice(24, 3).to_i
      oldcode = agehash[code]
      oldcode =~ / and /
      oldage = $`.to_i
      oldcount = $'.to_i
      newage = oldage + age
      newcount = oldcount + 1
      agehash[code] = newage.to_s + " and " + newcount.to_s
    end
  end
  seer_file.close
end
agehash.each_pair do
  |key, value|
  next if (value.eql?("0 and 0"))
  next if (codecount[key] < 80)
  number = codecount[key]
  number = "000000" + number.to_s
  number = number.slice(-6,6)
  value =~ / and /
  totalage = $`.to_i
  count = $'.to_i
  if (count > 0)
    average = (totalage / count).to_f
    average = average.floor.to_s
    outarray.push(average + " " + number + " " + dictionary[key])
  end
```

```
end
puts outarray.sort.join("\n")
exit
```

27.2 Analysis

The results of the exercise confirmed our prediction. Every type of cervical precancer had an average age of occurrence that was earlier than the average age of occurrence of every cervical cancer (Figure 27.1). Moreover, minimally invasive cancer of the cervix,

Avg. Age at occurrence		Diagnosis
Precancers		
34	049174	carcinoma in situ, nos
34	051911	squamous cell carcinoma in situ, nos
35	000359	sq. cell carcinoma, lg. cell, non-ker., in situ
37	000313	sq. cell carcinoma, keratinizing, nos, in situ
39	001348	adenocarcinoma in situ
39	018564	squamous intraepithelial neoplasia, grade iii
Microinvasive cancer		
41	003262	sq. cell carcinoma, micro–invasive
Carcinoma		
43	000130	adenocarcinoma, endocervical type
48	000093	large cell carcinoma, nos
48	001353	adenosquamous carcinoma
49	003189	sq. cell carcinoma, lg. cell, non–ker.
50	000272	endometrioid carcinoma
50	002573	carcinoma, nos
51	000105	sq. cell carcinoma, sm. cell, non–ker.
51	000253	small cell carcinoma, nos
51	000407	papillary adenocarcinoma, nos
51	002773	sq. cell carcinoma, keratinizing, nos
51	004268	adenocarcinoma, nos
52	000239	mucinous adenocarcinoma
52	019249	squamous cell carcinoma, nos
54	000098	papillary squamous cell carcinoma
54	000227	clear cell adenocarcinoma, nos
56	000142	mucin–producing adenocarcinoma
59	000304	neoplasm, malignant

Abbreviations: nos = not otherwise specified, sq = squamous, ker = keratinizing

Figure. 27.1 The average age of occurrence of the different types of precancers, and cancers occurring in the uterine cervix. The data indicates that the average age of occurrence of the precancers precedes the average age of occurrence of the cancers.

the earliest invasive cancer following cervical precancer and preceding the deeply invasive cancer of the cervix, occurred at an age intermediate between precancers and invasive cancers.

The precancers occurred with peak age in the mid to late 30s. The cancers occurred with peak ages in the mid 40s to mid 50s. This means that, on average, it takes years, possibly a decade or more, for precancers to develop into cancers.

Once again, an important medical question can be answered in a few lines of code, if we have access to a large, curated, and well-annotated data set, such as SEER.

Exercises

1. In this chapter's script, we columnated the output lines by front-padding variables with zeros, and than extracting a fixed length of the rightmost characters of the numeric entries. We could have saved ourselves a few lines of code by formatting the output lines with a printf statement. Using Perl, Python, or Ruby, modify the script to produce a neat column output, using a printf statement.

2. Precancers, unlike cancers, have a tendency to regress (disappear over time). In fact, precancer regression is thought to be a more frequent event than precancer progression (toward cancer). If this were true, we would expect to encounter many more cervical precancers than cancers in the SEER data sets. In this chapter's script, we did not actually tally the number of cervical precancers and cancers. Using Perl, Python, or Ruby, modify the script to capture and display the number of cases of each cervical precancer and cancer. Which lesions have the greatest frequency of occurrence in the SEER data records: precancers or cancers?

3. Cervical cancer in women is not the only type of cancer that has identifiable precursor lesions that is recorded in the SEER data records. The refractory anemias are thought to occur progressively over time (i.e., refractory anemia precedes refractory anemia with sideroblasts, which precedes the development of refractory anemia with excess blasts, which precedes the development of refractory anemia with excess blasts in transformation.

   Here are the ICD-O codes and terms for refractory anemia and related lesions:

   99803 Refractory anemia
   99823 Refractory anemia with sideroblasts
   99833 Refractory anemia with excess blasts
   99843 Refract. anemia with excess blasts in transformation
   99853 Refractory cytopenia with multilineage dysplasia
   99863 Myelodysplastic syndr. with 5q deletion syndrome
   99873 Therapy-related myelodysplastic syndrome, NOS
   99893 Myelodysplastic syndrome, NOS

   Using Perl, Python, or Ruby, write a script (you can modify the script in this chapter if you wish), that parses through all of the SEER cases, determines the average age of occurrence of each type of the lesions related to refractory anemia, and lists them, in order of ascending age of average occurrence of the lesion.

4. When we analyzed the average age of occurrence of cervical precancer, we did not need to perform a separate analysis of men and of women (because men do not develop carcinoma of the uterine cervix). Both men and women may develop refractory anemia. Using Perl, Python, or Ruby, write a script that determines the number of occurrences of refractory anemia, and related lesions, in men, and in women, and determines the average age of occurrence of each type of related lesion, in men and in women.

Epilogue for Healthcare Professionals and Medical Scientists

If you want to make your own important contributions to medical science using large data sets, you will need to acquire an intimate intellectual relationship with the data, and you will need to do some of your own programming to fully examine the data. If you leave all of your programming tasks to other people, you will never learn the true value of your data. If you follow a few simple suggestions, your data-intensive life will be productive and enjoyable.

Learn One or More Open Source Programming Languages

Students and healthcare professionals often ask themselves whether it is worth the pain and effort to learn another programming language. Those who use several programming languages stand to benefit in several ways.

1. When you rewrite a script in another language, you often discover mistakes in the original script. The most difficult mistakes to find are the ones that do not stop script execution. If you run three equivalent scripts, in three languages, and you find that each script produces a slightly different output, you know that something is not working as you had thought in one of your scripts. The process of reconciling this discrepancy, so that all three scripts produce the same output, produces better scripts, in all three languages.

2. Books are usually written with examples in one language only. Someday, you will find a book that covers an area of great interest to you, but you won't understand the lingo. People often choose a programming language for its popularity in their field. This is somewhat short-sighted. Today's hot language may be tomorrow's dud. Sometimes, the best book in your field was written

a decade ago for a language that was wildly popular at the time, but is now shunned by your contemporaries. By learning Perl, Python, and Ruby, you gain some perspective on the different ways a programming language can express common algorithmic tactics (conditionals, loops, assignments, ranges, references, data structures, etc.). You will find that by learning several languages, you can comfortably read almost any book that contains lines of programming code written for any language.

3. Learning several programming languages has the same advantage as learning several spoken languages. You are likely to meet someone whose primary language is different from your own, but with whom you share a secondary language. A shared language facilitates group efforts. Many C and Java programmers know several additional languages. It's likely that one of those languages will be Perl, Python, or Ruby.

Don't Agonize Over Which Language You Should Choose

Voltaire wrote, "The best is the enemy of the good." It seems that Perl, Python, and Ruby programmers are natural enemies, all jostling for the title of "best." This book demonstrates that, for the common computational tasks, it really doesn't make any difference which language is best. If you were a professional programmer, working with a team of programmers, analysts, and engineers under a tight deadline to produce a complex software application, the selection of a programming language might be important. But this is not the case here, and any language will serve you well.

Learn Algorithms

An algorithm is a step-by-step description of a procedure that solves a problem. I like to think of algorithms as perfect machines. They can be understood completely. If used correctly, they always work. They never grow old, never wear out, and never require maintenance. They take up no space and can be stored forever. They cost nothing and everyone can use them. Every process in the world that works reliably can be described with one or more algorithms.

It is important to distinguish computational algorithms from computer software applications. A software application is an assemblage of common algorithms, implemented in a programming language and tethered to a graphic user interface. Good computer programmers master hundreds of different algorithms. In well-written computer applications, the algorithms work together seamlessly to produce a desired functionality. Most people who use computer software applications are oblivious to the contained algorithms.

By learning basic algorithms, you gain a deeper knowledge of your field, and a comprehensive set of tools that can be used within any programming language that you choose to study.

Unless You Are a Professional Programmer, Relax and Enjoy Being a Newbie

After a few minutes of instruction, you can learn the rudiments of chess, and you can begin to play the game. You can spend the rest of your life trying to master the game. The same is true for Perl, Python, and Ruby programming. Luckily, most of the scripts that you will need in your professional life can be written with a very shallow skill set: open a file, read the lines of a file, look for a pattern in the file, make a substitution, extract a string, store information in a data structure, add information, count items, perform a numeric operation, and display the contents of a data structure. These basic elements of programming are easy to perform.

Developing a project, asking a good question, obtaining complete and accurate data, finding good co-workers, obtaining funding; these will always be the most difficult aspects of your professional life.

Do Not Delegate Simple Programming Tasks to Others

We cannot do everything for ourselves. In society, we often delegate tasks to trusted individuals who have specialized skills. We trust surgeons to remove our appendix when it is inflamed, dentists to fill our cavities, builders to construct our homes, educators to teach our children. Though it may seem absurd for healthcare professionals and medical scientists to do their own programming; it is necessary, just the same. The reason is that aside from the development of large applications (word processors, spreadsheets, databases, Web browsers, e-mail clients), most professional computational tasks are short but highly individualized operations. Most nonprogrammers do not have a programmer at their beck and call, willing to interrupt their work to listen to your very detailed request for a very small job. If you could find a programmer, what are the odds that they will understand how to use the data sources that are important for your project? Will they understand the words and concepts that are basic to your professional work? How will you compensate the programmer? Will you need to write a request for proposal, and will you need to select a contractor from among a list of responders? How much will you be willing to pay for an effort that, in the end, could have been completed with a few lines of code?

In my own experience, I am constantly appalled by the money and time invested in programming efforts that could have been achieved in a few hours by anyone with a little working knowledge of Perl, Python, or Ruby. Because many programming efforts require compliance with a set of specifications in a contract, on-the-fly changes in the project may be difficult or impossible to achieve. Unfortunately for you, writing a new program is like conducting a new experiment. You are constantly discovering that your initial assumptions were wrong, and you need to rethink your plans. In many cases, the program that eventually satisfies your needs is quite different than the program you originally requested. It is often the case that applications developed by professional programmers according to the terms of a contract do not provide the functionality that is ultimately required.

Many of the computational tasks that you will face in your professional life cannot be delegated. You will either do them for yourself, or they will not get done.

Break Complex Tasks into Simple Methods and Algorithms

Are you old enough to remember *The Jetsons*? This fabled cartoon show aired in the 1962–1963 TV season, and featured a futuristic family whose morning ablutions were co-opted by a mechanism that performed the following services quickly and efficiently: waking, washing, dressing, grooming, feeding, and depositing family members into the rocket-car. Nearly 50 years later, our simplest tasks of living lack any serious automation. Why? We manage to get out the door every morning under our own steam, or with the assistance of a few small devices (coffee maker, toaster, electric toothbrush), and we don't want a massive, complex device to control every step in the process. Humans excel at small, connected tasks that, in the aggregate, compose our lives.

You will find that any task in the field of biomedical informatics, no matter how daunting, can be broken into simple methods. Learning how to break a project into small tasks is itself an important life-skill. Once you've mastered the simple methods and algorithms in this book, you'll start seeing complex problems as a series of small problems that you will solve, confidently and eagerly.

Write Fast Scripts

I know from experience that the fastest way to kill any software project is to write a slow script. The following is a fictional example, loosely based on a real-life example.

I am informed by my co-workers that they have prototyped an application that will autocode medical reports. "How fast will the autocoder operate?" Nobody knows. The next day they return with an answer: 500 bytes per second. "What is that, in terms of the number of reports per second?" More confusion. The next day I learn that the average report is 1000 bytes. So a typical report can be automatically autocoded in 2 seconds. The team has 10 million reports on file, so that means that the autocoder can do its job in 20 million seconds. There are 604,800 seconds in a week. It would require over 8 months to do the job.

The team goes back to the drawing board. By improving the program, and by distributing the workload among a bank of computers, they have improved the prototype to the point that it can autocode 10 million reports in one week. The team is happy. They can start the job, go about their business, and return a week later to find the complete, autocoded output.

The plan is put into effect. A week later, when the output is reviewed, it is obvious that there is a flaw in the program. Many terms in the text are not provided with codes … something to do with an unexpected use of phrase modifiers in the reports that escaped the term matching subroutine. A correction is made, followed by another week-long test of the autocoder. This is followed by the discovery of additional problems with the matching algorithms, based on the inclusion of unexpected characters, misspellings, inappropriate separators, and a host of glitches. Each discovery prolongs the agony. The team decides to test the prototype on a smaller number of cases. This seems to work well, but the tests that worked well on a small sample of reports failed against larger samplings.

Finally, way over schedule, we have a fully autocoded set of reports. When we are asked to add 3 million reports, provided by a new hospital in the consortium, we proceed with confidence. Unfortunately, our confidence is not based on a realistic premise. The program fails miserably on the new reports, which were written in a style and format that evaded many of the autocoder's parsing routines.

More months pass. We finally produce an output that everyone could live with, executed in under a week. Unfortunately, we are told that the medical nomenclature that we had used was unsuitable. The administration has decided to switch to another standard vocabulary recommended by the U.S. government as part of an effort to standardize healthcare information. We need to start over, from scratch.

The scenario is always the same. Large data sets require fast software. You cannot improve, modify, repeat, or adjust to new conditions when your software is slow. When you think about it, the software you most enjoy is the software that responds instantly to user input.

It is often best to have short scripts that quickly (in a few seconds or less) parse through large data sets to produce an expected output. This is true, even if it means that you will need to use several different scripts, in tandem, to produce the final output you desire. Being able to inspect output by steps permits you to catch systemic errors and to assign those errors to one of a small set of subroutines.

For many of my projects, I develop a list of short, fast scripts that I employ in a certain order. I check whether the output from one step is ready to be used as the input for the next step.

When the project ends, I have an output file, but I do not have an application. There really isn't any need. Instead of having a deliverable software product, I end with a speedy set of small scripts. I have found that this method allows me to finish projects quickly and adapt to minor or major changes in the objectives of the project.

Concentrate on the Questions, Not the Answers

Analyses of large data sets most often produce somewhat tentative observations that yield more questions than answers. You always need to ask yourself whether the data set was built on faulty or inaccurate assumptions about reality, or whether there were systemic flaws in the way that the data was collected. Under the best circumstances, epidemiologic data yields statistical associations, without providing any proof of causal mechanisms. The astute healthcare programmer develops a new set of questions from every observation, and develops innovative methods to pursue those questions.

Appendix

How to Acquire Ruby

Ruby is a free, open source programming language that can be downloaded from multiple Web sites.

Linux and Windows® users can download Ruby from

http://rubyforge.org/frs/?group_id=167

How to Acquire Perl

Perl is distributed with most Linux operating system packages.

CPAN (Comprehensive Perl Archive Network) is the source for Perl and Perl modules:

http://www.cpan.org/

Windows users may find it convenient to use ActiveState's free Perl installation, available at

http://www.activestate.com/

The ActiveState installation provides access to the ActiveState Perl Package Manager, a quick way to install publicly available Perl modules.

How to Acquire Python

Python can be acquired at

http://www.python.org/download/releases/

How to Acquire RMagick

In Ruby, images are displayed using RMagick and Tk. RMagick is Ruby's interface to ImageMagick, a free software library for manipulating images. Tk is a free language for creating GUIs (graphic user interfaces). Tk employs widgets (small windows within the Tk window) for input and display structures.

After you have installed ImageMagick, RMagick, and Tk onto your computer, you can "require" them into your Ruby scripts and create applications that create, modify, evaluate, and display images. All three applications are available at no cost for users of Windows or Linux/Unix operating systems. Ample instruction is available at the Web sites listed later. Here are some suggestions for Windows users:

1. Go to the RubyForge site:

   http://rubyforge.org/frs/?group_id=12&release_id=8170

   This page has a combined win32 binary package for RMagick and ImageMagick Pick the binary that is appropriate for your version of Ruby.
   I use Ruby 1.8.4, so I chose the following binary:

   rmagick-1.13.0-IM-6.2.9-0-win32.zip 12.39 MB

2. Download the binary (zip file) and expand it.
   This produces the following subdirectory:

   rmagick-1.13.0-IM-6.2.9-0-win32

   The subdirectory contains a group of files:

   ImageMagick-6.2.9-0-Q8-windows-dll.exe
   readme-rmagick.html
   readme-rmagick.txt
   readme.html
   rmagick-1.13.0-win32.gem

3. Run the ImageMagick .exe file, and it will guide you through its installation.
4. After ImageMagick is installed, you can install the RMagick gem file by invoking Ruby's gem tool with an install command followed by the name of the gem file (add the full path to the gem file if you are not installing from its current subdirectory).

   c:\ruby>gem install rmagick-1.13.0-win32.gem

5. All the information you need to start using RMagick from within your own Ruby scripts is found at

   http://www.simplesystems.org/RMagick/doc/

Then install Tcl/Tk by visiting ActiveState and downloading the Activebinary for Windows users.

With these installations, you can write Ruby scripts that use and display images.

How to Acquire SQLite

SQLite is an extremely popular implementation of an SQL database. SQLite source code is in the public domain, and the application software is easily obtained and installed. SQLite permits users to create an SQL database on their own computer. Interfaces to SQLite have been written for many popular programming languages.

For Perl:

Perl's DBD::SQLite module includes the entire SQLite library, and the Perl interface to the library.

Information for the DBD::SQLite module is available at

http://search.cpan.org/~msergeant/DBD-SQLite-0.31/lib/DBD/SQLite.pm

The module can be downloaded and installed in one step from the ActiveState ppm manager, under the module name DBD-SQLite2.

For Python:

Pysqlite is the Python interface to SQLite. It includes the SQLite database software and the Python interface in a single distribution available from

http://code.google.com/p/pysqlite/downloads/list

The distribution can be built from source code, for Linux users, or installed as a precompiled binary (.exe) file for Windows users. Windows users should select the version of pysqlite that corresponds to the version of Python that resides on their own computer. The precompiled binary contains its own wizard installation.

A usage guide to Python's SQL interface is available at

http://koeritz.com/docs/python-pysqlite2/usage-guide.html

For Ruby:

Ruby users must first install SQLite on their computer, and then install the Ruby interface to SQLite, available as a Ruby gem.

To acquire SQLite, go to the SQL public download page (www.sqlite.org/download.html), and choose a download file appropriate for your computer's operating system. For Windows users, there are precompiled binaries. After downloading the precompiled binary for Windows, unzip the file and deposit the .dll in your ruby script subdirectory.

Next, install the Ruby gem that supports Ruby's interface to SQLite, calling the gem from your C prompt.

c:\>gem install sqlite3-ruby -v=1.2.3

You may be asked to select a gem appropriate for your system:

Select which gem to install for your platform (i386-mswin32)
1. sqlite3-ruby 1.2.3 (ruby)
2. sqlite3-ruby 1.2.3 (x86-mingw32)
3. sqlite3-ruby 1.2.3 (mswin32)
4. Cancel installation

Windows users should select item 3.

A usage guide to Ruby's SQL interface is available at

http://sqlite-ruby.rubyforge.org/sqlite3/faq.html

How to Acquire the Public Data Files Used in This Book

1. Medical Subject Headings (MeSH)
   The download site is

   http://www.nlm.nih.gov/mesh/filelist.html

   The file used in various scripts throughout this book is "d2009.bin", referred to as the ASCII MeSH download file. It is about 28 MB in length and contains over 25,000 MeSH records. A typical MeSH record is shown in Chapter 5, Figure 5.1.

2. The International Classification of Disease
   Let us start with the each10.txt file, available by anonymous ftp from the ftp.cdc.gov Web server at

   /pub/Health_Statistics/NCHS/Publications/ICD10/each10.txt

   Here are the first few lines of this file:

   A00Cholera
   A00.0Cholera due to Vibrio cholerae 01, biovar cholerae
   A00.1Cholera due to Vibrio cholerae 01, biovar el tor
   A00.9Cholera, unspecified
   A01Typhoid and paratyphoid fevers
   A01.0Typhoid fever
   A01.1Paratyphoid fever A
   A01.2Paratyphoid fever B
   A01.3Paratyphoid fever C

A01.4Paratyphoid fever, unspecified
A02Other salmonella infections
A02.0Salmonella gastroenteritis

3. International Classification of Disease Oncology Codes and Terms

The ICD-O (Oncology) file is available at

http://seer.cancer.gov/icd-o-3/sitetype.icdo3.d08152007.pdf (Figure A.1)

In several of the projects in this book, we use icdo3.txt, a plain-text reduction of the publicly available pdf file obtained at

http://www.julesberman.info/book/icdo3.txt

The ICD-O contains names of neoplasms. It was prepared by the SEER program to cover the Oncology (i.e., cancer) terms and codes recommended in the ICD by the World Health Organization, and referred to as version 3 of the oncology dictionary (ICDO-3). It contains 9,769 terms and codes.

8021/3 Carcinoma, anaplastic type, NOS
8022/3 Pleomoic carcinoma

The SEER files contain 5-digit codes, equivalent to the ICD-O codes, but with the "/" removed. For the SEER projects in this book, codes and terms are extracted from the icdo3.txt file, and the "/" is stripped from each term.

4. Data Dictionary for CDC Mortality Files
The data dictionary file is available by anonymous ftp from ftp.cdc.gov at the following subdirectory:

/pub/Health_Statistics/NCHS/Dataset_Documentation/mortality/
    Mort99doc.pdf

5. CDC Mortality Files
The files that we will use can be downloaded by anonymous ftp from the CDC server (ftp.cdc.gov)

The 1999 Mortality File

ftp server: ftp.cdc.gov
path: /pub/health_statistics/nchs/datasets/mortality
file: mort99us.dat (1,058,532,982 bytes)

2002 and 2004 Mortality Files

ftp server: ftp.cdc.gov
path: /pub/health_statistics/nchs/datasets/dvs/mortality
file: mort02us.dat (1,081,483,832 bytes)
file: mort04us.dat (1,176,686,000 bytes)

ICD-0-3 SEER SITE/HISTOLOGY VALIDATION LIST AUGUST 15, 2007

Most comparisons can be made to the three-digit histology code but a four-histology comparison is required wherever an '!' appears to the left of the three-digit histology name.
The Site/Type edit edits each morphology at the four-digit level.

LIP	C000-C006,C008-C009		
NEOPLASM		800	8000/3	Neoplasm, malignant

			8001/3	Tumor cells, malignant
			8002/3	Malignant tumor, small cell type
			8003/3	Malignant tumor, giant cell type
			8004/3	Malignant tumor, spindle cell type
			8005/3	Malignant tumor, clear cell type

CARCINOMA, NOS		801	8010/2	Carcinoma in situ, NOS
			8010/3	Carcinoma, NOS
			8011/3	Epithelioma, malignant
			8012/3	Large cell carcinoma, NOS
			8013/3	Large cell neuroendocrine carcinoma
			8014/3	Large cell carcinoma with rhabdoid phenotype
			8015/3	Glassy cell carcinoma

CARCINOMA, UNDIFF, NOS		802	8020/3	Carcinoma, undifferentiated type, NOS
			8021/3	Carcinoma, anaplstic type, NOS
			8022/3	Pleomorphic carcinoma

GIANT & SPINDLE CELL CARCINOMA		803	8030/3	Giant cell and spindle cell carcinoma
			8031/3	Giant cell carcinoma
			8032/3	Spindle cell carcinoma
			8033/3	Pseudosarcomatous carcinoma

Figure A.1   A sample page from the ICD-Oncology file.

The 1996 Mortality File (combines ICD9 and ICD10 encoded data)

ftp server: ftp.cdc.gov
path: /pub/health_statistics/nchs/datasets/comparability/icd9_icd10
file: ICD9_ICD10_comparability_public_use_ASCII.ZIP (130,240,471
 bytes)

In this book, the expanded ICD9 ICD10 comparability public use file was renamed to conform with the file names for the other mortality files: mort96us.dat (1,601,884,492 bytes)

The 1996 mortality file stores line-record ICD10 codes and terms in a different byte location than that used in the 1999, 2002, and 2004 mortality files. The data dictionaries for the 1996 mortality files contain the key to the byte locations of data in the 1996 file.

6. Online Mendelian Inheritance in Man (OMIM)
The compressed OMIM file (which exceeds 100 MB in length, uncompressed) is available for download by anonymous ftp from

ftp server: ftp.ncbi.nih.gov
path: /repository/omim/
file: omim.txt.Z

7. SEER data files
To get the SEER public use data files, you must first complete a data access request available at

http://seer.cancer.gov/data/request.html

SEER sends you a username and password that you will need to access the data files. The data is available on a DVD or by direct Internet download.

At the time that this book was written, the most recent SEER data covered 1973–2006, in the following release file:

04/15/2009 04:46 PM 223,935,710 SEER_1973_2006_TEXTDATA.
 d04152009.exe

The decompressed file contains a data dictionary (pdf file) and a subdirectory with the data files that are used in this book (Figure A.2):

04/14/2009 01:50 PM 153,783,644 BREAST.TXT
04/14/2009 01:50 PM 116,050,746 COLRECT.TXT
04/14/2009 01:50 PM 71,956,724 DIGOTHR.TXT
04/14/2009 01:50 PM 98,253,484 FEMGEN.TXT
04/14/2009 01:50 PM 75,934,488 LYMYLEUK.TXT

COMPUTER RECORD FORMAT

Item Name	Applicable Years	NAACCR Item #	Positions	Length
Patient ID number		20	01-08	8
Registry ID		40	09-18	10
Marital Status at DX		150	19-19	1
Race/Ethnicity		160	20-21	2
Spanish/Hispanic Origin		190	22-22	1
NHIA Derived Hispanic Origin		191	23-23	1
Sex		220	24-24	1
Age at diagnosis		230	25-27	3
Year of Birth		240	28-31	4
Birth Place		250	32-34	3
Sequence Number--Central		380	35-36	2
Month of diagnosis		390	37-38	2
Year of diagnosis		390	39-42	4
Primary Site		400	43-46	4
Laterality		410	47-47	1
Histology (92-00) ICD-O-2		420	48-51	4
Behavior (92-00) ICD-O-2		430	52-52	1
Histologic Type ICD-O-3		522	53-56	4
Behavior Code ICD-O-3		523	57-57	1
Grade		440	58-58	1
Diagnostic Confirmation		490	59-59	1
Type of Reporting Source		500	60-60	1
EOD—Tumor Size	1988-2003	780	61-63	3
EOD—Extension	1988-2003	790	64-65	2
EOD—Extension Prost Path	1985-2003	800	66-67	2
EOD—Lymph Node Involv	1988-2003	810	68-68	1
Regional Nodes Positive	1988+	820	69-70	2
Regional Nodes Examined	1988+	830	71-72	2
EOD—Old 13 Digit	1973-1982	840	73-85	13
EOD—Old 2 Digit	1973-1982	850	86-87	2
EOD—Old 4 Digit	1983-1987	860	88-91	4
Coding System for EOD	1973-2003	870	92-92	1
Tumor Marker 1	1990-2003	1150	93-93	1
Tumor Marker 2	1990-2003	1160	94-94	1
Tumor Marker 3	1998-2003	1170	95-95	1
CS Tumor Size	2004+	2800	96-98	3
CS Extension	2004+	2810	99-100	2
CS Lymph Nodes	2004+	2830	101-102	2
CS Mets at Dx	2004+	2850	103-104	2
CS Site-Specific Factor 1	2004+	2880	105-107	3
CS Site-Specific Factor 2	2004+	2890	108-110	3

Figure A.2 Part of the SEER data dictionary, describing bytes 1 through 110 of each SEER record.

04/14/2009 01:50 PM 128,405,116 MALEGEN.TXT
04/14/2009 01:50 PM 141,117,522 OTHER.TXT
04/14/2009 01:50 PM 136,211,152 RESPIR.TXT
04/14/2009 01:50 PM 63,487,816 URINARY.TXT

These files contain over 3.7 million records. Each record is a line of the file, and consists of 264 alphanumeric characters. A data dictionary provides the byte location of the various field values contained in each record. For the examples in this book, we will be using primarily age and diagnosis fields.

8. Topography codes (also called anatomic codes) used by SEER and by the WHO (World Health Organization) are available at

http://www.ncri.ie/data.cgi/html/icdo2sites.shtml

The first few codes are

C000 External lip upper
C001 External lip lower
C002 External lip NOS
C003 Upper lip, mucosa
C004 Lower lip, mucosa
C005 Mucosa lip, NOS
C006 Commissure lip
C008 Overlapping lesion of lip
C019 Base of tongue, NOS

9. State Codes are available as Item 2 in the CDC mortality documentation, available at: ftp.cdc.gov/pub/Health_Statistics/NCHS/Dataset_Documentation/mortality/Mort99doc.pdf. A copy of the state codes is also available at www.julesberman.info/book/cdc_States.txt.

10. Year 2000 Census Data
The MR(31)-CO.txt is a public domain file, 65,046,291 bytes in length, available from the U.S. Census Bureau at

http://www.census.gov/popest/archives/files/MR-CO.txt

A Web page providing a general description of this file is available at

http://www.census.gov/popest/archives/files/MRSF-01-US1.html

And a data dictionary for the file is available at

http://www.census.gov/popest/archives/files/MRSF-01-US1.pdf

This file is very important because it contains detailed year 2000 census and ethnicity data for states and counties. The year 2000 is used as the standard population against which other population-based epidemiologic data is adjusted.

11. Year 2000 United States Age Population File
The file can be downloaded from

http://www.census.gov/popest/archives/EST90INTERCENSAL/US-EST90INT-07/US-EST90INT-07-2000.csv

A copy of the file can be downloaded from

http://www.julesberman.info/book/censuage.txt

12. The Developmental Lineage Classification and Taxonomy of Neoplasms is an open source computer-parsable data set that can be used to organize, collect, merge, share, analyze, understand, develop, and test hypotheses, and discover new information related to neoplasia.

     The Classification and Taxonomy contains about 6,000 classified types of neoplasms and over 135,000 neoplasm names. It is the largest cancer nomenclature in existence and has been described in the following citation:

Berman JJ. Tumor classification: molecular analysis meets Aristotle. BMC Cancer, BMC Cancer 2004, 4:10.

The Neoplasm Classification is available in XML, RDF, and flat-file formats, available at

http://www.julesberman.info/devclass.htm

Other Publicly Available Files, Data Sets, and Utilities

1. GZIP
   GZIP compresses and decompresses files. Files with a .gz or .Z suffix usually require GZIP decompression (with the companion GUNZIP utility)
   Information and downloads are available at

http://www.gzip.org/

2. 7-ZIP
   7-ZIP is open source software that can be obtained at

http://www.7-zip.org/

7-ZIP can compress, archive, decompress and de-archive using a variety of popular archive and compression formats.

3. Medical Word List
   An open list of about 50,000 medical words is available at the OpenMedSpel site:

http://www.e-medtools.com/openmedspel.html

4. DICOM images
   Many DICOM images are available at the following site:

ftp://ftp.erl.wustl.edu/pub/dicom/images/version3/RSNA95/

Index

circumstances. On the other side of the implementation coin, freedom to translate legislative mandates into administrative action can lead to a wide range of interpretations from minor modifications to goal distortion and displacement.

Whether a regulatory agency, for example, adopts a general rule by which to decide whether violations have occurred or adopts a case-by-case approach has significant consequences for how the agency conducts its work as well as for the outcomes of the agency's decisions. Use of a general principle will produce more consistent and uniform decisions compared to a case-by-case approach; on the other hand, a general principle as a decision criterion is less sensitive to individual variation in cases. Here one can see the dilemma discretion poses for public administrators: which of several important democratic values should be served?

Political Context and Implementation

Scholars of policy implementation have long observed the importance of economic, political, cultural, and social change on the success or failure of a public policy (Van Meter & Van Horn, 1975). Certainly, economic, cultural, and social trends impact not just the demand and design of public policy, but also make it easier or harder to achieve a policy's goals. A program to assist displaced workers by retraining them as computer programmers may well fail to achieve its purpose of reducing unemployment because employers decide to "outsource" jobs for basic computer programmers to overseas locations. The unexpected arrival of non-English-speaking refugees may overwhelm an agency's ability to assist other indigent persons in the locality and may prompt the agency to hire interpreters. Police officers may arrest a parent for what appears to be a case of child abuse when, in fact, the spots on the child's back are the result of warm stones, a folk medicine practice common in some cultures.

Political conditions constitute the most immediate and important parameter on policy implementation. Changes in the partisan affiliation of elected executives may lead to new and different orders to career administrators that direct them to enforce the law more or less vigorously. The emergence of a new interest group or social movement (e.g., consumer protection) may alter an agency's established relationship with other interest groups (e.g., manufacturers) such that the agency is caught between contending parties. As a consequence, agency managers may be forced to seek compromises that alter the agency's practices.

Changes in the political context of a policy may also occur because of shifts in public opinion. When the change in public opinion is large and coalesces around a relatively clear position, as noted earlier in this chapter, elected policy makers are likely to respond. However, it is less certain that bureaucrats respond to shifts in public opinion. Stillman (1996, p. 100) claims that "generally public opinion is too transitory and amorphous to have much influence on the typical public agency." But Stillman also says that he is not suggesting that public opinion never influences administrative actions. Agencies may be sensitive to general public opinion because a poor public image can affect the agency's standing with elected officials. Over time, agencies may alter their practices to increase their support from the general public, as the Army Corps of Engineers did in transforming itself from an agency perceived as a detriment to the environment to an agency with a substantial commitment to environmental protection (Mazmanian & Nienaber, 1979). Studies of variation in the political cultures of American states also indicate that citizens in different parts of the country hold different ideological orientations toward government and politics including the roles of public administrators (Flentje, 2000). These different cultural expectations establish parameters of what constitutes acceptable styles of state and local administration. On balance, public opinion does have some effect on policy implementation, and civil servants do not necessarily behave in accordance with the Kafkaesque image some have of bureaucrats.

POLICY DESIGN, BUREAUCRACY, AND DEMOCRACY

Surprising to some, many democratic theorists do not discuss or go into much detail about the place and role of bureaucracy. Concern over bureaucracy, when the topic is broached, focuses on the question of whether bureaucratic officials will faithfully follow the policy directives of legislatures and elected executives. That administrators will use their resources to alter public policy and thus weaken popular sovereignty is the principal democratic fear. This concern of course, is an important one in public administration. The debates over how best to ensure the accountability of bureaucrats have a long and distinguished history, and numerous means have been proposed by which to control civil servants (Gormley & Balla, 2004, pp. 10–11).

But another question about the use of bureaucracy as an instrument of popular sovereignty also bears on the place and role of bureaucracy in a democratic government—in what ways does the public expect and empower public administrators to act? To be sure, the public does not and will not support a "rogue" bureaucracy, but the public does expect administrative agencies to act and to act in ways preferred by the public, for example, respond to a disaster in an expeditious and effective manner. Considine (2005, pp. 191–192) argues that "policy-making always involves a dual structure." By this he means every policy design has an "instrumental dimension" of decisions, programs, and outcomes as well as a "developmental dimension" of norms and relationships. That is, policy choices not only seek to achieve particular goals and objectives, but the goals and objectives contained in policy decisions communicate expectations about the future condition of the community or nation (e.g., less crime). As Schneider and Ingram instruct (1997, pp. 140–141): "Policy teaches lessons about what groups people belong to, the characteristics of groups with which people identify, what they deserve from government, and what is expected from them . . . whether the problems of the target population are legitimate ones for government attention, what kind of game politics is (public-spirited or the pursuit of private interests), and who usually wins." The developmental dimension of policy design also includes expectations about the behavior of bureaucrats and their relationships with the policy's targets and the general public—for example, the IRS is expected to collect taxes and apprehend tax cheaters, but to do so in a way that is not burdensome or overly intrusive to ordinary citizens.

Smith and Ingram (2002, p. 566) decry the "blind spot" policy studies have for the relationship between policy tool choices and the condition of democracy. But policy design also includes more than just the "instrument of action" by which the policy goal is to be achieved; policy design creates an official role for administrators and among these many and diverse roles are: agent, expert, guardian, leader, mediator, partner, servant, and steward. It is important to understand that the role assigned to administrators establishes more than just a function to be performed, it also includes norms about the interaction between citizens and administrators. The interaction of tool choice with role expectation affects agency performance and policy outcome, and crucially defines the standards of accountability to which the policy and its administrators are held (Posner, 2002). Rule *by* the people can be protected from "unfaithful" actions by imposing controls on unelected officials, but it is important to understand that rule *for* the people can be expanded through the roles and relationships embodied in a policy's design.